Gene Traders

Biotechnology, World Trade, and the Globalization of Hunger

EDITED BY BRIAN TOKAR

TOWARD FREEDOM

Toward Freedom
Burlington, Vermont

Toward Freedom
PO Box 468
Burlington, VT 05402
www.TowardFreedom.com

To purchase copies or request one for review, e-mail Info@TowardFreedom.com or call (802) 657-3733.

ISBN 0-9746935-1-0

Cover illustration: Mary Azarian
Cover design: The Laughing Bear Associates
Book production: Anne Linton, MacWorks
Index: Dian Mueller

Printed in the US by Capitol City Press

® GCIU

Contents

Introduction:
Resisting Biotechnology and Global Injustice

BRIAN TOKAR

I N NOVEMBER OF 1999, TENS OF THOUSANDS OF PEOPLE — union members, students, environmentalists, parents, and many others — converged in Seattle to nonviolently challenge the biennial ministerial meeting of the World Trade Organization (WTO). While media accounts focused on the Seattle police's unprecedented brutality toward demonstrators — and on the actions of small numbers who chose to fight back — the importance of Seattle went far beyond the debates surrounding the police and protesters' tactics.

WTO delegates representing countries from around the world were newly emboldened by this first tangible sign that people in the US were ready to join millions of others worldwide in opposing the Brave New World of corporate globalism. The delegates were able to successfully resist US pressure to launch a new "Millennium Round" of global trade talks, aimed at greatly expanding the purview of the WTO, and the meeting ended with no substantive agreement. Similar scenarios would be played out at future meetings as well. Twenty-two countries of the global South (what the US press, politicians, and many academics continue to call the "developing world") walked out of WTO negotiations in Cancun, Mexico, in 2003, and the US conceded to a considerably narrowed vision of its proposed Free Trade Area of the Americas later that same year.

The Seattle protests embraced a breathtaking scope of social, economic, and environmental issues, equal in breadth to the vast array of policies, affecting all the world's peoples, over which the WTO was seeking dominion. In the face of worldwide resistance to expanding global "free market" policies — also known as "neoliberalism"[1] — global elites had come to see international trade agreements as a means to impose policies that would never withstand the scrutiny of democratic debate in most countries. The most notable of these measures, codified in the founding documents of the WTO, condemned public policies aimed at protecting people and the environment as barriers to global trade — barriers that could be contested, and ultimately overturned,

1

by a secretive body of international trade officials.

The people who gathered in Seattle represented a unique convergence of labor and environmental concerns in the US: the much-touted alliance of "teamsters and turtles." The endangered Pacific sea turtle had become a powerful symbol of the environmentally destructive nature of the WTO, after trade authorities overturned a US law requiring that shrimp exporters use nets especially designed to avoid trapping these turtles. Many environmentalists in Seattle displayed costumes and signs celebrating the sea turtle as one of the principle non-human victims of economic globalization.

Another lasting symbol of the Seattle protests was the monarch butterfly. Six months earlier, just as the biotechnology industry lobby was holding its annual convention in Seattle amid small but determined protests, scientists at Cornell University made headlines worldwide with their announcement that pollen from a widely-planted variety of genetically engineered (GE) corn was lethal to monarch butterflies in their larval (caterpillar) stage.[2] The splendid monarch became an international symbol of opposition to GE foods and crops, and banners and costumes bearing the images of butterflies, along with turtles, were seen everywhere in the teeming Seattle crowds. With the WTO debating restrictions on national food safety measures, as well as rules mandating the patenting of life forms in all member countries, it was clear to many that genetic engineering was a cornerstone of corporate-driven globalization. This relatively new technology had become a central means by which global corporations, as well as financial institutions like the WTO, were expanding their control over the food supply, personal health, and environmental security of people throughout the world.

As the images of Seattle passed into history, people in Europe, Asia, and other parts of the world continued as before to confront the biotechnology industry as a central manifestation of global corporate control. In the United States, however, biotech and global justice activists followed rather divergent paths. US global justice campaigners, empowered by the outcome of Seattle, felt for the first time that it was possible to directly confront the core institutions of global capitalism. Economic issues rose to the forefront and environmental concerns, including genetic engineering, were seen as less central to the core mission of challenging the WTO and other global institutions.

For some US activists, GE issues were too technical and too specific, a distraction from the heady opportunity to challenge global capitalism as a whole. Others were put off, sometimes justifiably, by the common framing of genetic engineering as a consumer issue. For example, efforts to require labeling of genetically engineered food products appealed mainly to those affluent enough to pay more for safer foods. "GE-Free" was in danger of becoming an elite niche market, as was already happening with organic foods, much to the chagrin of the organic movement's visionary pioneers. At the same time, high-profile environmental organizations had been positioning themselves for over a decade as participants in the Washington, DC, policy establishment rather than the voices of a social movement.[3] This legacy further contributed to the frequent marginalization of environmental concerns within the emerging global justice movement. While some campaigners continued to emphasize the fundamental links between genetic engineering and globalization, particularly around the series of "Biodevastation" events that usually coincide with the annual biotech industry conventions in various cities,[4] the two movements proceeded for some years along largely independent paths.

Meanwhile, throughout much of Europe, Asia, Latin America, and Africa, popular movements around global economics, genetic engineering, and local food sovereignty evolved along a far more unified course. Wherever global corporations threatened the integrity of local food cultures and agricultural practices, wherever the WTO, World Bank, and International Monetary Fund were supporting policies that forced traditional food growers off their land, wherever threats to the integrity of living forests, rivers, and coastlines directly impacted traditional agrarian communities, resistance to biotechnology grew hand in hand with resistance to globalization.

In Britain, hundreds of people participated in the sabotage of field tests of genetically engineered crop varieties. Sometimes in the dark of night, but increasingly in broad daylight, with the participation of people of all ages and backgrounds, colorful processions marched on GE test plots, pulled the experimental crops out of the ground, and bagged them as hazardous waste.[5] In several high-profile cases, people were tried and acquitted for doing such actions. Opposition to genetic engineering peaked at well over 90 percent of the British population, and millions of concerned people came to view Prime Minister Tony Blair's advocacy of GMOs (genetically modified organisms) as part of the same misguided policy as his dogged support for the 2003 US invasion and occupation of Iraq.

In France, a large movement of dissident farmers, organized under the national Confédération Paysanne, confronted GMOs along with other symbols of global corporate domination. In 1998, a group of farmers entered a Novartis warehouse containing five tons of genetically engineered corn and destroyed the crop by spraying it with water hoses and fire extinguishers. A year later, Roquefort cheese producers from the south of the country blockaded McDonald's fast food shops with tractor loads of rotten fruit and manure, and sometimes filled the restaurants with flocks of live chickens and turkeys. They were responding to a series of new retaliatory tariffs imposed by the US against European goods, including specialty cheeses, in retaliation for Europe's refusal to import American beef raised with a host of synthetic hormones. José Bové, a prominent spokesperson for the farmers' confederation, became an international hero when he was imprisoned for participating in the systematic dismantling of a McDonald's under construction in his hometown of Millau.[6] He began traveling to global summits and trade meetings throughout the world to protest GMOs and other impositions of US-based agribusiness corporations.

In India, farmers organized in even greater numbers against the increasing corporate control over seeds. In the mid-1990s, as commercial GE crops were about to be introduced worldwide, Monsanto bought a 26 percent share of one of India's largest seed suppliers, the Maharashtra Hybrid Seed Company (now MAHYCO). Facing the contamination of traditional crop varieties by GMOs, as well as intense pressure from the WTO to allow the patenting of living organisms under Indian law, hundreds of thousands of farmers joined demonstrations, and a large coalition of groups launched a nationwide "Monsanto Quit India" campaign in 1998, on the anniversary of the day Gandhi urged the British to finally "Quit India."[7]

In the southern Indian state of Karnataka, farmers burned several test plots of genetically engineered cotton and declared their intention to "Cremate Monsanto." In the central state of Andhra Pradesh, local farmers uprooted a GE cotton crop, and then successfully petitioned their state government to ban further trials and uproot several more test plots. The Indian Supreme Court responded favorably to a petition to consider whether the planting of genetically engineered crops violates fundamental constitutional rights. While indebtedness resulting from an increasingly unsustainable reliance on agricultural machinery and chemicals has driven thousands of Indian farmers to suicide, organizations throughout the country are working to revive the use of traditional seed varieties and wean entire villages from their dependence on chemical agriculture.[8]

In Brazil, the Movimento dos Trabalhadores Rurais Sem-Terra (MST), or Landless Rural Workers' Movement, has rallied well over a million displaced people to occupy and reclaim idled agricultural estates. In a country where less than three percent of the population owns two-thirds of the farmland, and some 60 percent of that land remains idle, the MST's strategy of land occupation, resettlement, and development of new agricultural cooperatives has brought new hope to people suffering under the full weight of neoliberal economic policies. The success of the organization's alternative economic models has contributed to its popularity and rapid growth across the vast Brazilian countryside.[9]

In 2001, as tens of thousands of people gathered for the first World Social Forum in the southern Brazilian city of Porto Alegre, some 1200 residents of MST-organized communities stormed a regional Monsanto research facility, pulled experimental GE corn and soybean plants out of the ground, and set up hammocks and mattresses for a long-term occupation.[10] José Bové and other farmer activists from around the world, who were gathered in Brazil for the Social Forum, traveled to the site to join the MST protest. A 2003 decision by the popular Brazilian president Luiz Inácio Lula da Silva to temporarily permit the growing of GE soybeans in Brazil led to a protracted political standoff, as the country risked losing its status as the world's largest supplier of non-GE soybeans.

In the summer of 2002, southern African countries facing a threat of growing food shortages shocked the world by refusing a US offer of food aid in the form of GE corn. African leaders were concerned about the health consequences for their people — who often rely on corn for upwards of 90 percent of their daily calories — as well as the imminent threat of contamination of local corn varieties if any of the aid corn were to be planted. Several countries ultimately accepted the GE corn, though many required that the corn first be ground into meal to avoid inadvertent planting. Only Zambia, where officials launched a nationwide public debate on GE food, stood firm in refusing this aid. The GE food aid debate in Zambia, along with many other stories from around the world, is recounted in detail in this book.

The convergence of biotechnology and corporate globalism once again came into sharp focus in the North in May 2003, when the US announced that it would seek WTO action against Europe's *de facto* moratorium on approving new GE crop varieties. Literally speaking, Europe as a whole had never legislated such a moratorium. However, no new GE varieties had been approved for import into Europe since 1998, individual countries were raising increasingly determined obstacles to GMOs, and there was a virtual freeze on growing GE crops in Europe, except for small amounts in Spain and the Netherlands. Bush administration officials proclaimed that an intervention at the WTO level was necessary to protect American farmers, seeking to paint a populist face on their typically aggressive intervention on behalf of US agribusiness interests.

Almost immediately, people on both sides of the Atlantic raised objections to this move. At the seventh annual Biodevastation gathering — held that year in Monsanto's home town of St. Louis, and highlighting the essential links between genetic engineering, corporate control, and environmental racism — the participants approved a declaration opposing the WTO intervention. The statement supported European resistance to GMOs, condemned the Bush administration's action, endorsed precautionary measures to protect public health and the environment, and demanded that commercial developers of hazardous chemicals and GE technologies be held fully liable for the damage they cause.[11] In the lead-up to the 2003 WTO meeting in Cancun, this evolved into a Global Citizen's GMO Challenge, focusing on the need for precaution, people's right to protect their food choices and intervene in international proceedings, and the prevalence of alternatives to GMOs. The Challenge, along with a European counterpart initiated by Friends of the Earth, attracted the endorsement of hundreds of organizations from around the world.[12]

Meanwhile, during the Winter of 2002, activists on both US coasts had discovered plans by the Department of Agriculture (USDA) to host an international summit of agriculture and trade ministers in Sacramento, California, during June 2003. Following a tentative step toward reviving global trade talks in the Qatari city of Doha in 2001, the US administration sought to bring the world's trade and agriculture ministers to one of the political centers of US agribusiness just three months prior to the WTO talks in Cancun. The Sacramento event was billed as an agricultural technology expo where the wonders of California's agriculture would be demonstrated to the world. For global justice activists, however, it was apparent that the US government's main agenda was to do everything possible to prevent another Seattle-style collapse at the upcoming WTO negotiations.

In early 2003, activists based in California and Vermont created the Sacramento Mobilization for Food Sovereignty, Democracy and Justice and began organizing to bring people to Sacramento during the USDA's ministerial conference and technology expo. The mobilization attracted the support of 130 organizations — including the United Farm Workers' union and the international farmers' movement Via Campesina — and brought close to 5000 people into the streets of California's state capital. What USDA officials had planned as a quiet, behind-the-scenes official meeting, out of public view, sparked a major protest against biotechnology, industrial agriculture, and the US government's preparations for Cancun, exposing a wide array of issues to a renewed level of scrutiny and public attention.

The events in Sacramento highlighted the recent US appeal to the WTO to overturn European restrictions on GE crops, the refusal of GE food aid by Zambia, and the Bush administration's subsequent attempt to link AIDS relief to countries' acceptance of GMOs, along with a host of other immediate issues. Antiwar activists from throughout California denounced the unprecedented rise in biotech corporate subsidies in the name of combating "bioterrorism." Others worked to highlight movements for food sovereignty and self reliance from around the world, as well as from California's own inner cities. There was a standing-room-only teach-in at a local university, several days of direct actions focused on corporate agribusiness and the WTO, and a huge rally and parade, which became a massive festival of resistance to biotechnology, industrial agriculture, globalization, and US imperial designs around the world. Some people locked themselves to trees in support of a local community garden that had recently been evicted, well-equipped climbers scaled a huge DNA sculpture to protest research on genetically engineered trees at a nearby university, and a group of Midwestern farmers dumped GE corn on the sidewalk just as USDA officials were cutting the ceremonial ribbons to open their official meetings. Other activists engaged in several days of attempted street blockades in the vicinity of the Sacramento Convention Center. Sacramento set a creative and inspiring example for people simultaneously concerned about globalization and genetic engineering, and revealed how agribusiness and biotechnology play a far more central role in the global designs of the US administration than many people had previously realized. [13]

Why has the technology of genetic engineering inspired such widespread and determined opposition throughout the world? First and foremost, it is the most visible means by which global corporations are consolidating their control over our food and health, a condition that has evolved over several decades. As growing environmental awareness during the 1970s and Eighties aroused fears on the part of agribusiness executives that the age of chemical agriculture could be coming to an end, they came to see their possible salvation in the brand new technology of gene splicing, or "recombinant DNA." But the story began even earlier than that.

With pesticide sales increasing a hundredfold since the 1940s, companies like Monsanto, DuPont, and Dow had acquired tremendous leverage to advise farmers, and ultimately determine how most of our food is grown. After DDT was banned in the US in 1972, and prohibitions on other toxic pesticides soon followed, the companies anxiously sought other means to sustain this control. Genetic engineering appeared to be the solution. By shifting their technological interventions right into the genetic makeup of seeds, companies could make farmers highly dependent on new, patented seed varieties, as well as on the particular chemicals with which those seeds were "designed" to grow. In the late 1990s, Monsanto alone spent at least $8 billion acquiring several of the leading commercial seed companies in the US and around the world.[14] The company now spends $10 million each year bringing lawsuits against farmers who are believed to be growing Monsanto's GE crop varieties in ways that violate the company's mandatory licensing agreements and patents.[15]

Popular concerns around genetic engineering reach far beyond those directly involved in agriculture, however. Since the introduction of the first commercial GE crop varieties in 1996, independent scientists have discovered a host of disturbing human health and environmental consequences. For example, proteins that cause allergic reactions are passed from one organism to another through genetic engineering, and GE foods may be introducing brand new allergens into our diet, as with the StarLink variety of GE corn that was never approved for human consumption, and which forced the recall of over 300 contaminated US food products in 2000 and 2001 (see Chapter 1).

The spread of antibiotic resistance is another unexpected consequence of genetic engineering. Scientists insert genes for antibiotic resistance (so-called "marker genes") as a means to identify the tiny fraction of cells that are "successfully" genetically engineered in their laboratories, and the DNA conferring antibiotic resistance can ultimately be passed along to disease-causing bacteria. A senior scientist working in a Scottish genetic engineering lab discovered even more disturbing effects of GE food on laboratory rats, including a depressed immune response, inflammation of their intestines, and dramatic alterations in the sizes and weights of many vital organs.[16] Overall, very few laboratories around the world have the funds to carry out experiments on GE food effects. Agribusiness corporations have tremendous economic and political leverage over the priorities in agricultural research, and have aggressively intervened to discredit dissenting scientists. Still, every year brings further confirmation of the suspicion that GE foods are harmful to human health.

The discovery in 1999 of the deadly effects of pollen from genetically engineered corn on immature monarch butterflies — with an almost 50 percent mortality rate for caterpillars that were directly exposed to the altered pollen — offered dramatic evidence for the environmental consequences of genetic engineering. While industry spokespeople insist that the famous monarch studies have since been refuted, the real story is far more ambiguous and disturbing.[17] Harm to other beneficial insects, the threat of "superweeds" resistant to multiple chemical herbicides, genetic contamination from engineered trees and fish, and other present and future hazards have all contributed to a growing level of environmental concern.[18]

The threat of environmental contamination is most serious in tropical regions, where wild relatives of many common food crops can be found, but genetic contamination of our basic staple foods is a concern for everyone. A 2004 study by the Union of Concerned Scientists demonstrated that two-thirds of common varieties of conventional soybean, corn, and canola seed used by Midwestern farmers may be contaminated with small but significant amounts of DNA from several widely used GE crop varieties.[19] GMOs represent forms of living pollution that are able to grow and reproduce freely, spreading genetic instability and disruption to a host of related organisms in a manner that may never be completely reversed.

Genetic engineering also raises a host of ethical, political, and even cultural and aesthetic concerns. Tampering with the underlying structures of life — by means that overturn the very processes of genetic regulation that help make us who we are — raises alarm for reasons that are sometimes very personal. In the short term, genetic engineering can violate religious strictures against consuming certain foods or combinations of foods, especially where they cannot be clearly identified. But it also raises much wider concerns about the integrity and very meaning of life on earth, whether viewed in a religious or secular context.

The technology also opens the door to far more disturbing interventions in the medium- and long-term future, including the growing practice — thus far experimental — of genetically engineering animals and, perhaps before long, the genetic engineering of human beings. Some biotech advocates celebrate the arrival of a "post-human" future in which "we" (or, more likely, powerful institutions outside of our control) directly intervene in the genetic makeup of our children. Modern biotechnology has the means to reduce all of life on earth to a set of patentable gene sequences and other commodities, available to be bought, sold, and traded in the commercial marketplace. The announcement by South Korean scientists in early 2004 that they had successfully created 30 cloned human embryos raised these concerns to a newly heightened level of alarm.[20]

In the US today, industry-funded advocacy groups like the Biotechnology Industry Organization, CropLife America, and the Council for Biotechnology Information spend well over $50 million per year trying to convince a skeptical public that genetic engineering has important benefits and is worthy of support.[21] Even local campaigns to label GE foods or halt the use of GMOs have faced massively funded opposition. Biotech companies spent at least $5 million to defeat a statewide GE food labeling initiative in Oregon in 2002, and proponents of a 2004 referendum to prohibit the raising of engineered plants and animals in Mendocino County, California, had to overcome over $700,000 in industry PR to win passage of their proposed ban.[22] The industry's benefactors have spent over $100 million to develop the so-called "golden" vitamin A rice, which is aggressively promoted as evidence that GMOs will "feed the world." This despite the observation that a person would have to consume massive quantities of rice every day to satisfy their requirement for vitamin A, and would still require sufficient body fat for the vitamin A precursor in the rice to be properly metabolized (see Chapter 7). This led author and *New York Times* commentator Michael Pollan to accurately describe vitamin A rice as "the world's first purely rhetorical technology."[23]

The myth that biotechnology is poised to "feed the world" is perhaps the most pervasive of the numerous false hopes that the developers of this technology have aroused. Indeed, people around the world have had to work overtime to dispel this myth, and expose the many ways genetically engineered agriculture is undermining food security and the survival of land-based peoples. In 2001, for example, the world renowned scientist and activist Vandana Shiva denounced the unapproved planting of almost 3000 acres of GE cotton in western India — by a MAHYCO-Monsanto joint venture — as an illegal act of bioterrorism, with no measures in place to protect the region's 130 indigenous cotton varieties from contamination.[24] In 2003, an alliance of indigenous and *campesino* organizations throughout Mexico discovered that the transgenic contamination of ancient traditional corn varieties — which had aroused a worldwide scientific debate when initially discovered two years earlier — had spread to at least 33 communities in nine Mexican states, despite a continuing prohibition against growing GMOs in Mexico.[25] This case will be discussed in some detail in Chapter 1 of this book.

Feeding the world is only possible if people can exercise the fundamental ability and right to feed themselves, and genetic contamination from GMOs threatens this in myriad new and profound ways, as we will see. This is especially clear when contrasted with the alternatives. While GE crop varieties have a neutral-at-best effect on crop yields, research combining indigenous knowledge with the scientific application of organic crop improvement methods has led to far more consistent benefits.[26] Crop rotations, natural soil amendments, and detailed studies of the life cycles of various "pest" species — a focus of worldwide research until the dawn of the pesticide era that immediately followed World War II — have far more to offer the world's farmers than the endless cutting and splicing of DNA in pursuit of new commercially viable and patentable "products."

The last time a convergence of new technology and aggressive economic policy so sweepingly impacted the food choices and economic status of people around the world was during the so-called Green Revolution in agriculture, which began in the 1950s and continued through the Seventies. With the completion of the post-World War II reconstruction of Europe, US-based foundations, government agencies, and international institutions like the World Bank all turned their sights toward the so-called developing world. With the final demise of European colonialism during the 1950s and early Sixties, newly independent nations in Africa, Asia, and Latin America were exploring a variety of development models and building new social and economic institutions. In the geopolitical climate shaped by the Cold War between the US and the Soviet Union, newly developing countries often became proxies for one or the other side in that conflict. To counter the growing appeal of socialist-oriented development models in the so-called Third World, US-based institutions sought to bolster post-colonial economies while simultaneously assuring the participation of these emerging nations in the US-dominated global market economy.

The Green Revolution began with the introduction of new, higher-yielding corn varieties in Mexico, but reached its fruition in India and other south Asian countries, where attractive packages of loans, equipment, and technology were offered to farmers who would make the transition toward more intensive market-oriented production. The "high-yielding" varieties of corn, rice, and other staple grains that emerged from new US-funded research institutes throughout the world (see Chapter 3) produced higher yields for two main reasons. First, they were usually dwarf varieties, producing more seeds on smaller stems. Second, researchers were actively breeding and selecting varieties that showed much greater responsiveness to mechanical cultivation and applications of chemical fertilizers. Often these varieties were more susceptible to pest infestations and required increasing applications of chemical insecticides as well. Governments that supported the shift toward mechanized, chemical-intensive agriculture were rewarded by the US government and the World Bank with favorable loan conditions and other forms of economic aid, while farmers were won over with offers of free seed and other benefits. Soon, they found themselves increasingly dependent on the use of costly equipment and chemicals.

In many regions, agricultural productivity increased dramatically at the outset, but within a decade began to level off and even decline.

Agricultural regions faced severe water shortages, as new World Bank-financed irrigation systems exhausted water supplies and dramatically lowered water tables. Farmers growing dwarf varieties of rice, and other crops that produced less straw, no longer had sufficient fodder to feed their livestock, and grain monocultures replaced the diverse traditional cropping systems that had fed and sustained people on their traditional lands, often since prehistoric times. The Green Revolution enriched some farmers, and enabled them to greatly expand their land holdings, while others were plunged into debt and dependency. Entire communities were forced to migrate to overcrowded urban areas where they had to earn cash to buy their food. The few prospered, while entire regions became mired in scarcity, dislocation, and internecine violence.

Vandana Shiva writes:

> The very meaning of agriculture was transformed with the introduction of the western Green Revolution paradigm. It was no longer an activity that worked towards a careful maintenance of nature's capital in fertile soils and provided society with food and nutrition. It became an activity aimed primarily at the production of agricultural commodities for profit.[27]

In a subsequent detailed study of the effects of these policies on the northern Indian Punjab region, she added:

> The Green Revolution was based on the assumption that technology is a superior substitute for nature, and hence a means of producing growth, unconstrained by nature's limits.... The reduction in availability of fertile land and genetic diversity of crops as a result of the Green Revolution practices indicates that at the ecological level, the Green Revolution produced scarcity, not abundance.[28]

Devinder Sharma, a contributor to this volume, has described the Green Revolution as a failure, a "frightening scenario" of chemical dependence and inequality that has left the land and the people "gasping for breath."[29] Today's biotech advocates often describe their technology as the harbinger of a "new Green Revolution." With the presumed "benefits" of genetically engineered agriculture far more speculative than those of earlier agricultural technologies, it is essential that their plans continue to be viewed with skepticism and determined opposition.

The essays in this book offer a broad and comprehensive view of the many ways that global trade agreements, international financial institutions, governments, and agribusiness corporations are furthering the expansion of genetically engineered agriculture around the world. Each of the authors presents a distinct case study, showing how the interplay of trade policy, "development" politics, and biotechnology has increased dependency and hunger, while compromising the survival of traditional farmers and their communities. The first three essays focus on the major international financial institutions and trade agreements: NAFTA, the Free Trade Area of the Americas (FTAA), the WTO, and World Bank, documenting in detail the ways each of these institutions has promoted the biotech agenda in agriculture. S'ra DeSantis begins by outlining the specific hazards of GMOs and how "free trade" policies have led to the widespread contamination of ancient, indigenous corn varieties in Mexico and Central America. Aziz Choudry focuses his analysis on the World Trade Organization, describing how various trade agreements under its jurisdiction have compelled countries around the world to accept GMOs and allow the patenting of seeds and other life forms. My own chapter focuses on the World Bank, documenting how the Bank's policies since the Green Revolution have systematically introduced and promoted GE agriculture in the far corners of the so-called "developing world."

The last four essays focus on particular regions of the global South and how the politics of hunger and "development" have furthered the expansion of GE agriculture. Lawrence Tsimese of Ghana begins by framing the growing debate over GMOs in Africa and its relationship to agricultural sustainability. Shiri Pasternak focuses on the institutions of US food aid policy, particularly the US Agency for International Development and the World Food Program, and how US policies have subsidized agribusiness expansion under the veil of humanitarianism. Dr. Mwananyanda Mbikusita Lewanika recounts the dramatic story of Zambia's rejection of GE food aid in 2002, resisting overwhelming US and international pressure. Finally, Devinder Sharma shows how biotech interests have sought to exploit epidemics of hunger and malnutrition in his native India, while he proposes a different kind of development agenda that supports farmers and promotes agricultural self-sufficiency.

A word on terminology is in order before proceeding. Readers will find that various terms — genetic engineering (GE), genetic modification (GM), and transgenic, as well as the popular acronym GMO, for genetically modified organism — are used interchangeably in this book. At the outset of the biotech era in the 1970s, genetic engineering was the widely accepted term to describe the brand new technology of gene splicing recombinant DNA. In Europe, people referred to both "genetic engineering" and "genetic manipulation." In 1988, the European Commission adopted the term "genetically modified" in its first policy directive on the release of engineered organisms, a euphemism clearly crafted to moderate public fears of this still-new technology.[30] Of course "engineering" can imply a degree of certainty and predictability that this technology has not begun to approach, while in much of the world the term GMO has come to represent all the potential horrors of genetic manipulation. "Transgenic," while still somewhat esoteric, is perhaps the most scientifically accurate description of this technology, emphasizing the unique character of transferred genes. I hope readers who are somewhat new to these terms will become increasingly familiar with them as you read on. Other terms used frequently by critics of globalization are explained in context the first time they arise, either in the text or in the endnotes. Abbreviations are spelled out in the Index.

Finally, it is my pleasure to thank all the wonderful people without whose dedication this book would never have found its way to print. First I'd like to offer my tremendous gratitude to all the contributors, who in every case condensed many years of research, experience, and wisdom into their chapters and, in many instances, held up to the pressure of multiple edits under tight deadlines. Next, I offer my profound thanks to the inspired and dedicated staff and editorial board of *Toward Freedom*, particularly Greg Guma, Carol Liu, Dian Mueller, and Chris Donnelly for their invaluable editorial insights and logistical support, as well as *TF* publisher Robin Lloyd, who has kept the dream of a world moving "toward freedom" alive these many years. I also want to thank my colleagues at the Institute for Social Ecology and the wider GE-Free Vermont coalition — particularly Dan Chodorkoff, Michael Cuba, Claudia Bagiackas, Bayard Littlefield, and Amy Shollenberger — for offering me some reprieve from other responsibilities this past Winter in order to focus on bringing this book to completion.

I also want to offer a special word of thanks to my colleagues Arthur Foelsche, Doyle Canning, and Brooke Lehman, who returned from Seattle in early December of 1999 with a gleam in their eyes, profound new insights about the links between globalization, biotechnology, and a host of interrelated social and ecological concerns, as well as a sense of urgency that movements in the US should evolve to better illustrate and embody those links. Their generation of activists has renewed the spark of hope and determination that social movements in the US were lacking for several decades, and rekindled the flame of awareness that a better world — founded on true equality and freedom — is both necessary *and* possible.

Earth Day, 2004

Control through Contamination:
Genetically Engineered Corn and Free Trade in Latin America

S'RA DESANTIS

THROUGH TRADE AGREEMENTS AND INTERNATIONAL FOOD AID, US-based agribusinesses are contaminating indigenous corn crops throughout Mexico with DNA from genetically modified organisms (GMOs). The apparent strategy of these corporations is to spread genetic contamination throughout the world through future free trade agreements that would force poorer countries to accept exports of genetically modified seeds and products. Transnational corporations, especially Monsanto, want to promote the belief that the world's food supply is thoroughly contaminated, so there can be no effective regulation or avoidance of GMOs. The North American Free Trade Agreement (NAFTA) has allowed the United States to dump millions of tons of corn into Mexico. Thirty to forty percent of this corn is genetically modified[1] and it has adulterated indigenous corn with its altered, or transgenic, DNA.

The contamination in Mexico serves as a prime example of how genetically engineered crops and free trade interface: together, they create the genetic pollution of corn in its center of origin. Trade agreements like the Free Trade Area of the Americas (FTAA), the Central American Free Trade Agreement (CAFTA), and several pending bilateral agreements, will secure dumping grounds for genetically engineered products, since numerous countries throughout the world continue to close their doors to such imports.

This chapter first describes the types of genetically engineered crops on the market and the dangers they pose to human health, the environment, and family farms. The focus then shifts to the specific case of contamination of indigenous corn in Mexico and how NAFTA was a major cause of this pollution. Finally, the chapter addresses two pending regional trade agreements and their potential to spread the contamination farther.

TYPES OF GENETICALLY MODIFIED CROPS

The corporate control of life has now reached the most intimate cellular level through the process of genetic engineering. In nature, bacteria and differing plant and animal species cannot share genetic material. But genetic engineering has eliminated these boundaries: Now scientists can exchange portions of DNA among animals, plants, fungi, and bacteria. These processes happen in laboratories usually controlled by corporations like Monsanto, Bayer, and Syngenta.

The process of genetic engineering is far from precise. It usually requires a "gene gun," an appropriate name for such a violent technology. Microscopic pellets of gold or tungsten are coated with DNA from one or several organisms. The gene gun fires these pellets into another organism's embryo cells at velocities up to 1400 feet per second. Some of the DNA from the pellets enters the nuclei of the host cells, randomly inserting foreign DNA into the host's genome. Often, thousands or millions of shots have to be fired to create a few "successfully" engineered plants.[2] Scientists who genetically alter organisms have barely begun to study the effects this haphazard genetic disruption produces in their altered creations. Nearly all of today's genetically engineered crop varieties are of two types: those that resist insects through the production of an insecticidal toxin from bacteria, and those that tolerate applications of a particular company's broad-spectrum herbicide.

INSECTICIDE-PRODUCING CROPS

GE crops that produce an insecticide within their cells are known as "insect-protected crops" (IPCs). This biotechnology term is misleading, however, since the plants will only be "protected" against certain insects that the insecticide intentionally or unintentionally targets. They are more appropriately named "insecticide-producing crops" because the insecticidal toxins are genetically engineered into every cell of the plant. The industry alleges that IPCs will benefit the environment because less chemical insecticide will be sprayed on plants when IPCs are cultivated. But the use of insecticides actually increases with IPCs, as the plants produce them internally, even as other insecticides continue to be sprayed topically.[3]

All of the commercially available IPCs use insecticidal proteins from "Bt," or *Bacillus thuringiensis,* a naturally occurring soil bacterium, which produces a series of protective proteins toxic to various insects. These bacteria, commonly applied to crops by organic farmers, produce toxins that are quickly broken down, and are only activated following ingestion by insects with alkaline digestive systems. GE plants containing the Bt genetic code, however, are capable of producing the toxin in its most biologically active form in every plant cell, and can produce up to 20 times as much toxin as natural Bt bacteria. Genetically engineered Bt crops currently on the market are corn, potatoes, and cotton, with dozens of others awaiting approval.[4]

Several negative effects have been associated with the use of IPCs, resulting from the continual production of activated Bt toxins. Bt pollen can harm beneficial insects, and the plants cross-pollinate with wild relatives or neighboring crops. Scientists worldwide have predicted that target pests will become resistant to Bt because not all insects will die from eating pesticidal crops; instead, some of these survivors will evolve to become resistant.[5]

Among the several beneficial insects harmed by pesticidal crops are lacewings, monarch butterflies,[6] and ladybugs.[7] Studies demonstrate that the lifespan and egg production of ladybugs are decreased, the survival rate of green lacewing larvae diminishes, and honeybees' foraging behavior and lifespan are reduced when these insects are exposed to Bt in its activated form.[8] Swiss scientists showed that lacewings die when they feed on larvae of European corn borers that have ingested Bt.[9] Ladybugs and lacewings are critical for natural pest control and honeybees are essential for the pollination of numerous crop plants. No one knows what effect the artificial Bt, also present in the roots, has on microorganisms or insects in the soil. Scientists based in Venezuela and the US found that the toxin secreted from Bt corn roots remained bound to soil particles in its active, lethal state for at least seven months.[10]

Problems with Bt have occurred higher up the food chain as well. Farmers in Iowa have reported that pigs fed with Bt corn have exhibited behavioral and physical signs of being pregnant even though they were not, rendering breeding impossible. The only practice that these farmers shared was that they all fed their animals Bt corn. When one farmer stopped feeding Bt corn to his pigs, they were able to become pregnant again.[11] In the past two years, more than 20 farmers have reported problems with false sow pregnancies evidently caused by Bt corn diets.[12] These dangers of Bt are only beginning to be documented; additional short- and long-term problems are sure to surface if the use of these crops continues to spread.

HERBICIDE-TOLERANT CROPS

Several corporations have genetically engineered crops to be resistant to herbicides. Glyphosate, the most commonly used broad-spectrum herbicide, kills all leafy plants, in contrast to more common plant family-specific herbicides. Roundup, Monsanto's brand of glyphosate, kills plants by inhibiting the enzyme functions required to synthesize certain amino acids. "Roundup Ready" crops are engineered through the use of bacterial and petunia genes to resist damage from Roundup, and are by far the most common type of GE crops grown today.

More than 75 percent of US soybeans, 65 percent of US cotton, and 10 percent of corn are Roundup Ready.[13] Monsanto's Roundup Ready retail crops include corn, sugar beets, canola, cotton, and soybeans; wheat, turf grass for golf courses, and alfalfa may be nearing commercial release.

One of Monsanto's main public relations claims is that genetic engineering will reduce herbicide use, despite the evidence from their Roundup Ready products. The application of various glyphosate varieties has increased by two and a half times since Roundup Ready crops were first introduced in 1996,[14] and total pesticide use has increased by an estimated 70 million pounds since 1996 from the use of herbicide tolerant crops.[15] Farmers can spray unlimited amounts of glyphosate on these resistant crops without killing them. The weeds around them will perish — until they, too, build up a resistance from Roundup exposure. As weed species are evolving tolerance to Roundup, farmers are using increasing quantities and concentrations of the herbicide, usually in tandem with other, even more toxic, chemical applications.[16]

Glyphosate herbicides cause many health and environmental complications. Mammals experience toxic reactions including convulsions and respiratory arrest. Genetic and reproductive damage has been found in many species exposed to the herbicide.[17] Some of Roundup's "inert" ingredients are also acutely toxic, resulting in gastrointestinal pain, vomiting, fluid in the lungs, pneumonia, and destruction of red blood cells, mucous membranes, and the upper respiratory tract.[18] Japanese researchers studying cases of suicide or attempted suicide from glyphosate ingestion determined that drinking little more than

3/4 of a cup is fatal,[19] and exposure to Roundup has also been associated with cancer and birth defects in humans.[20]

Glyphosate in the soil remains active and absorbs into soil components. Residues from the herbicide have been discovered on lettuce, carrots, and barley a year after Roundup was used in the same field.[21] Glyphosate kills not only plants, but also beneficial insects like parasitoid wasps, lacewings, and ladybugs, and damages earthworms and beneficial fungi, as well as fish and amphibians. Transgenic Roundup Ready canola has impaired bees' ability to recognize the scents of flowers, making it harder for them to find plants to pollinate.[22] Roundup also inhibits nitrogen fixation, an essential process for regenerating the soil.[23]

TERMINATOR TECHNOLOGY

The USDA, Monsanto, and other biotechnology corporations are developing a type of seed aptly dubbed "Terminator," genetically engineered with a suicide mechanism that prevents the next generation of seeds from germinating. Since these seeds cannot germinate, seed saving becomes impossible. Seed saving is crucial for food security worldwide since not everyone is willing or able to buy commercial seed.

Terminator technology exposes the hypocrisy of Monsanto's claim that genetically engineered crops will feed the world. The industry's PR campaigns also claim that Terminator will prevent cross-pollination of non-GE crops by altered varieties, thereby controlling genetic contamination from such plants as Bt corn. However, if Terminator crops *do* cross-pollinate with neighboring crops, these crops will likely inherit the Terminator trait and become sterile.[24] Terminator technology forces farmers to buy seeds from corporations every year, and severely undermines their ability and their right to save seed. Even though Terminator's release into the market has been delayed for the time being, biotechnology corporations are still developing it and hoping to prevail.

PHARMACROPS

GE crops are now being grown to produce pharmaceuticals and industrial enzymes. For example, Epicyte, based in San Diego, California, is producing GE corn that will contain a spermicide.[25] Biotechnology corporations prefer corn as a vessel to "grow" pharmaceuticals because it is relatively easy to genetically alter and simple to store, transport, and cultivate.[26] In July 2002, Friends of the Earth discovered over 300 secret field trials of pharmacrops, engineered to produce abortion-inducing chemicals, growth hormones, vaccines, and other drugs.[27] Some of the largest food companies in the US have lobbied for a zero-tolerance policy toward possible contamination of food crops by drug-producing varieties.[28]

In 2001, the Texas-based company, ProdiGene, contracted with farmers in Iowa and Nebraska to plant field trials of GE "pharma-corn" to produce a pig vaccine. In Nebraska the following year, the farmers cultivated soybeans, but residual ProdiGene corn seed from the year before sprouted and contaminated their harvest. Traces of pharma-corn were mixed in with 500 bushels of soybeans. These were sent to a grain elevator where they were mixed with 500,000 bushels. ProdiGene was forced to purchase and destroy all these soybeans for $2.8 million.[29] In Iowa, ProdiGene destroyed 155 acres of corn because it was believed to have cross-pollinated with a pharma-corn from the previous harvest season that sprouted in a test field.[30] Many other cases of contamination of field crops with pharmacrops have likely occurred but have not yet been detected.

HUMAN HEALTH AND GMOs

Almost all GE food crops contain antibiotic resistance genes, which are inserted to help scientists determine whether the new genetic material was successfully transferred. This process can introduce the antibiotic resistance trait into the environment, and into the bacteria that inhabit the human gut. These bacteria can then pass the trait on to pathogenic bacteria, also making them resistant to antibiotics. One of Novartis' (non syngenta) GE corn varieties was resistant to ampicillin, a common antibiotic for humans and other animals.[31]

GE foods can also transfer food allergies from one plant to another as genes corresponding to allergy-causing proteins are spliced across the species barrier. GE foods are also likely to activate additional, unknown allergens as genes from bacteria, viruses, and non-food plants produce proteins that have not previously been contained in our food. When a gene from another species makes its way into our food, the potential effects on the human bodies' are unpredictable. New chemicals toxic to humans could be formed, and the biochemical processes that maintain the integrity of our body's cells can be altered in unanticipated ways.

Scientists at the US Food and Drug Administration (FDA) have warned about the dangers of such latent toxins in GE foods. The GE "FlavrSavr" tomato, released in the mid-1990s and subsequently withdrawn, caused stomach lesions in laboratory rats. In 1990, 27 people died and 1500 became severely ill from L-tryptophan, a genetically engineered dietary supplement.[32] Genes from Brazil nuts were introduced into soybeans to improve their nutritional content, even though some people have fatal reactions to Brazil nuts. When blood samples from people with Brazil nut allergies demonstrated an allergic reaction in the presence of these GE soybeans, these soybeans were kept off the market. What would have happened, however, had these soybeans been allowed to enter the food supply, like so many other virtually untested GE varieties?

The massive contamination of the US corn supply by a GE variety called StarLink serves as one of the most striking examples of cross-pollination and mixing of seed lots. StarLink corn, genetically engineered by Aventis, contained the Bt bacterial protein Cry9C toxin, which human digestion cannot break down, and is classified as a likely human allergen. The US Environmental Protection Agency (EPA) approved StarLink only for animal feeds. In September 2000, activists at Friends of the Earth in the US discovered that Taco Bell's taco shells, and over 300 other name-brand products, were contaminated with StarLink corn. Eighty varieties of yellow corn seed were found to contain Cry9C, and the FDA and the US Centers for Disease Control investigated 40 cases of StarLink-induced allergies.[33] In October 2000, the USDA recalled 350,000 acres of StarLink corn planted in the US.[34] Some of the corn had been processed into taco shells at a *maquila* in Mexicali, Mexico, which also manufactures Sabritas brand Mexican snack foods. The US-bound taco shells were recalled from the shelves, but the Mexican Sabritas products were not.[35]

US federal inspectors found traces of StarLink in 10 percent of 110,000 tests conducted between November 2000 and April 2001,[36] even though only 0.4 percent of cultivated US corn was StarLink.[37] Cross-pollination and mixing of crops caused the discrepancy. Today, StarLink continues to be detected throughout the world. In South Korea, 55,000 tons of corn certified as StarLink-free tested positive.[38] Japan also discovered tainted US corn imports,[39] and food aid sent to Bolivia tested positive for traces of StarLink corn.[40] In 2003, three years after it was forcibly withdrawn from the market, this banned variety was still detected in one percent of corn samples submitted by growers and grain handlers.[41]

Since genetic engineering alters the constituents of food, many scientists assert that it also alters nutritional value. The FDA's Division of Food Chemistry and Technology and Food Contaminants Chemistry warned in 1992 that GE would reduce nutrient value. Dr. Arpad Pusztai, a specialist at Scotland's Rowett Research Institute, conducted a study to determine the health effects of potatoes genetically engineered to contain a lectin, or protective toxin, derived from snowdrops. The amounts of protein, sugars, and starch in the GE potatoes varied as much as 20 percent from non-engineered potatoes.[42]

Pusztai also evaluated the metabolism and organ development of rats fed GE potatoes with that of two control groups fed non-GE potatoes. The rats fed with GE potatoes showed significant changes in the size and weight of their intestines, pancreas, kidneys, liver, lungs, and brains. There was also evidence of intestinal inflammation and abnormal cell growth, as well as suppression of the immune system.[43] The rats that ate non-GE potatoes, even if they were sprayed with the lectin, did not exhibit any of the effects associated with GE potatoes.[44] Pusztai concluded that the effects were associated with the DNA constructs that these experimental GE potatoes share with other GE foods. After Pusztai went public with his results, Rowett officials attacked his work and fired him. The publicity from this case helped educate people throughout Europe about the potential dangers of GE foods.

THE LEGAL RISK TO FARMERS

Throughout the world, farmers save seed to cultivate the following year, stewarding their stock and developing new varieties. Intellectual property rights granted to seed manufacturers threaten farmers' and indigenous peoples' ability to save seed legally.

Farmers growing any of Monsanto's genetically engineered crop varieties sign a contract and pay a "technology fee" to use the patented seed. Anyone found cultivating such a crop without a signed contract can face legal action by Monsanto, whether or not the farmer knew that the crop contained patented genetic material. Hundreds of farmers throughout North America have been sued or threatened by Monsanto and have settled out of court.

Monsanto has established an entire department to enforce its licensing agreements and GE seed patents, with 75 employees and an annual budget of $10 million.[45] Monsanto sends inspectors to make sure farmers do not save seed, and has a hotline so people can anonymously turn in neighbors they suspect of illegally saving or using Monsanto GE seeds.

Percy Schmeiser, a farmer and seed saver from Saskatchewan, Canada, had his canola fields contaminated, either by cross-pollination or by seeds blown in from a neighboring farm. Acting as "gene police," Monsanto agents trespassed on his land and sampled his canola to look for their patented "Roundup Ready" gene. They sued Schmeiser for US$125,000 for illegally growing their Roundup-tolerant GE canola, even though Schmeiser never intended or wanted to cultivate it, and never sprayed his fields with Roundup to take advantage of the GE trait.[46] The Canadian Supreme Court heard Schmeiser's appeal in January 2004, and a decision is expected in the Summer or Fall.

THE CASE OF OAXACA: HOW FREE TRADE AND GMOS INTERFACE

One of the most daunting consequences of GE is contamination of neighboring crops and their wild relatives. Because they reproduce, GE plants can be classified as "living pollution." Once a GE crop enters the environment, its pollen — containing unique combinations of genetic traits whose long-term effects have not been studied — can be spread. The pollen is spread by wind and insects, contaminating other plants of the same species, and potentially reproducing the pollution indefinitely. The genetic contamination of indigenous corn varieties in Mexico serves as the most explicit example of how NAFTA's trade regime and GMOs interface. This catastrophe must serve as a warning call to the world.

Campesinos (peasant farmers) and corn have a symbiotic relationship. Neither could survive without the other. Corn provides *campesinos* with nutrition, economic livelihood, and a basis for many religious ceremonies. Mexican farmers maintain traditional varieties and breed new ones, acting as the stewards of corn and the protectors of biodiversity. There are over 20,000 varieties of corn in Mexico and Central America;[47] in southern and central Mexico alone, researchers have identified 5000 varieties.[48] Each has adapted to regional elevation, soil acidity, sun exposure, soil type, and rainfall. In 1998, the Mexican Congress passed a moratorium on the cultivation of genetically engineered corn to protect indigenous varieties from biotech-induced extinction.

David Quist, a graduate student at the University of California at Berkeley, and Ignacio Chapela, professor of Environmental Science, Policy and Management at the same institution, discovered that corn in the Mexican state of Oaxaca contained DNA from GMOs (transgenic DNA).[49] Their findings prompted two Mexican governmental agencies, the National Commission on Biodiversity (CONABIO) and the National Ecological Institute (INE), to sample indigenous corn from 20 communities in Oaxaca and two in the state of Puebla. They found that 95 percent of these communities (21 out of 22) had a contamination rate between one and 35 percent.[50] In total, GMOs polluted eight percent of the 1876 seedlings tested.[51] If such pollution exists where it is illegal to grow genetically engineered corn, what is the contamination rate in countries like the US and Canada where it is legal?

After the discovery of contamination in Oaxaca, indigenous and farming organizations conducted further tests on corn throughout the country.[52] Random samples were taken from rural communities where farmers save seed. They found contaminated indigenous corn in nine states: Chihuahua, Durango, Mexico State, Morelos, Oaxaca, Puebla, San Luis Potosi, Tlaxaca, and Veracruz. In all nine a significant percentage of the corn tested positive for Cry9C, the StarLink Bt-protein, approved in the US only for cattle feed. Other strains of Bt corn and genes from Monsanto's Roundup Ready corn were also detected in these states. One plant from Oaxaca contained genetic material from three different GE varieties.[53]

Despite these findings, the Mexican government said it would proceed with plans to lift its moratorium on experimental GE corn cultivation.[54] Industry scientists have pressured the government to rescind the moratorium, ostensibly so they can study issues such as gene flow. Looking at the tests already conducted, cross-pollination and genetic drift obviously occur. Unfortunately, such informal experiments have already occurred and continue daily in farmers' fields. Introducing more GE corn into the environment will only exacerbate the contamination rates.

As genetically modified corn cross-pollinates with indigenous varieties, the transgenic DNA could interfere with the expression of unique physical characteristics and genetic predispositions, making the indigenous corn less suitable for its particular environment. As indigenous varieties lose these qualities, yields will decrease and the *campesinos'* livelihoods will be further undermined. The development of new, locally appropriate, corn varieties is also threatened. Also, the genetically altered corn could cross with *teosinte*, the wild ancestor of corn, which grows in and around the edges of cornfields in southern Mexico.[55] The distinct genetic composition could be lost in *teosinte* and other relatives of corn as they cross-pollinate with genetically contaminated corn. With their extinction, a unique part of the earth's heritage would be lost forever.

GENETIC CONTAMINATION IN MEXICO: A NAFTA HORROR STORY

NAFTA, which allowed for unlimited imports of genetically modified US corn, directly caused the contamination in Mexico.[56] Before NAFTA, Mexico was practically self-sufficient in corn production; today, the country is a major importer. During NAFTA negotiations, a comparative advantage analysis was conducted in the three signatory countries — Canada, the US, and Mexico — to determine what each should produce for export. Mexico was not chosen to grow corn, even though it is the center of origin for this crop. Instead, the US was selected to produce corn and other basic grains, since it controls technologies like chemical fertilizers, pesticides, hybrid seeds, irrigation, and mechanized equipment. The analysis indicated that Mexico should cultivate labor-intensive horticultural crops such as flowers and winter vegetables, since it has a large, cheap labor force.[57] This analysis failed to consider the impacts this would have on the *campesino* way of life and corn biodiversity.

Additionally, this analysis overlooked the large government subsidies that US farmers receive, underscoring the contradictions of "free trade." US farmers on average receive 30 times more in subsidies than their Mexican counterparts,[58] with large-scale farmers receiving a disproportionate share of these subsidies. Medium-sized farmers in the US also depend on subsidies to survive; otherwise many more farms would go under. This dependence has been created by the falling prices of agricultural products on the global market, due to corporate concentration and inequitable trade policies.

NAFTA facilitated the further opening of the corn market in Mexico. The Agricultural Agreement of NAFTA, Chapter 7, eliminated all tariffs on agricultural goods either immediately or within a five-, ten-, or fifteen-year period. The agreement established a duty-free quota system or protection period for corn with a fifteen-year phase out. In 1994 — NAFTA's first year — 2.5 million metric tons of corn were permitted to enter Mexico from the US, tariff-free. The amount the US could export tariff-free was to increase 3 percent annually for 14 years, completing the 15-year phase-out period. Corn imports from the US surpassing the duty-free quota would be subject to a tariff. This phase-out period, however, did not last very long. In 1996, corn imports exceeded the quota by over three million tons and all tariffs were waived. In every subsequent year, the US surpassed the quota and tariffs were not applied. In the absence of tariffs, US corn exporters were given a green light to send more cheap, subsidized corn to Mexico.

Before NAFTA, the US exported approximately two million metric tons of corn annually to Mexico. In 2001, Mexico received 6.2 million tons of corn from the US.[59] An estimated 26 percent of the corn grown in the US in 2002 was genetically modified.[60] Both the European Union and Japan maintain restrictions on the import of GE foods, disproportionately increasing the quantity of GE grain the US exports to other countries. Hence, an estimated 30-40 percent of the corn exported from the US to Mexico is genetically engineered.[61]

Corn imports from the US were the primary direct source of the genetic contamination in Oaxaca. Diconsa, a state-run grain distributor, facilitated the dispersion of this GE corn. Diconsa delivers grain and other supplies to stores throughout Mexico's rural areas. According to Manuel Mérida at the Diconsa warehouse in Oaxaca City, 40 percent of the corn distributed in Oaxaca by Diconsa in 2001 originated in the US.[62] CONABIO and INE found 37 percent of the corn in the Diconsa warehouse in Ixtlán to be contaminated.[63] Nonetheless, a worker at the Ixtlán store reported that a Diconsa representative informed her there was no GMO corn in the store.[64] Another worker at the Guelatoa store was told, "GE corn is colored and Diconsa only sells white corn, so there is nothing to worry about."[65] There are no signs in the stores warning *campesinos* not to cultivate the Diconsa corn, even though it is highly tainted with transgenic DNA. Six workers at Diconsa stores throughout the Sierra Juárez stated that *campesinos* know not to plant Diconsa corn, and only cultivate their *criollo* (indigenous) varieties. However, two of 29 campesino families interviewed in January of 2002 admitted that they had experimented with Diconsa corn.[66]

Diconsa corn has made its way into the ground through many avenues. For centuries, farmers have conducted agricultural experiments. A few *campesinos* in the Sierra Juárez tried planting Diconsa corn, since they had never been informed of its dangers. It falls off trucks during delivery and grows by the roadside. Also, Diconsa corn that is used as livestock feed but is not completely consumed often ends up germinating. It eventually cross-pollinates with *criollo* varieties, and the contamination spreads. Other potential sources of contamination include illegal planting by transnational corporations, government distribution of GE seed, and international food aid. Ending imports alone will not stop this self-replicating pollution.

US corn is sold on the Mexican market for 25 percent less than its production cost,[67] driving down the price of local corn and undermining the *campesinos'* ability to make a living. Umberto Rosales, an engineer from Mexico's Secretariat for Agriculture, Livestock, Fisheries and Food (SAGARPA), declared that *campesinos* should produce higher yields of corn, cultivate a more profitable crop, or leave the land.[68] Trade-related economic conditions already impel over 600 *campesinos* to abandon farming every day.[69] Since NAFTA's implementation, more than 1.8 million rural-dwelling Mexicans have lost their jobs.[70] The US is now seeking to pass CAFTA and the FTAA, extending the contamination of corn throughout the Americas.

FREE TRADE: CONTAMINATING THE WORLD

President George W. Bush has witnessed defeat after defeat in his attempt to expand the US "free trade" agenda. Masses of people continue to descend on cities where trade negotiations are held to protest the trading away of people's livelihoods, health, and rights for the sake of profit. In September 2003, as thousands took to the streets of Cancun, 22 countries walked out of World Trade Organization (WTO) negotiations because the US and Europe refused to eliminate their agricultural subsidies and import tariffs. The US and its allies continue to demand that other countries stop protecting their agricultural sectors, while US subsidies to its own agribusiness corporations continue to rise. More than half of the 22 dissenting countries are also engaged in FTAA and/or CAFTA negotiations with the US, and have been subjected to a host of pressure tactics and threats from US Trade Representative Robert Zoellick and other officials.

Since 1994, the US has pressed for a free trade zone extending from northern Canada to Tierra del Fuego, expanding NAFTA and reinforcing the Monroe Doctrine. The latter policy, issued in 1823, declared Latin America and the Caribbean to be under the sphere of influence of the US, not Europe. The FTAA would encompass all countries in North, South, and Central America and the Caribbean with the exception of Cuba.

As an estimated 25,000 people converged on Miami in November 2003 to protest the FTAA, trade ministers signed a watered-down version of this agreement. Wanting to avoid another political embarrassment following Cancun, the Bush administration brokered a deal with Brazil to minimize potential conflicts. The FTAA version signed in Miami — known as "FTAA lite" or "FTAA *a la carte*" because it abandoned most of the original text — would allow countries to opt out of various agreements within the FTAA, such as those governing agriculture or services, or to choose their own commitment levels in each of the nine areas of negotiation.[71] This hemispheric agreement will permit the US to continue funneling enormous agricultural subsidies to corporate-run "farms," while allowing countries like Brazil to uphold restrictions on foreign investment and maintain their own laws on intellectual property rights. The amended FTAA faces a tough fight in the US Congress, especially from corporate lobbies objecting to the deletion of important provisions granting them access to markets and cheap labor.

Because the US has not been able to get what it wants in multilateral or hemispheric trade agreements, it is selecting countries for bilateral pacts, such as was recently signed with Chile. Other regional arrangements are also on the table. By making other countries fear exclusion from the trade arena, the US pressures them to open their borders to the "free" market. The US has announced plans to negotiate bilateral treaties with Bolivia, Colombia, the Dominican Republic, Ecuador, Panama, and Peru.[72] Venezuela was the only Andean country not invited to join the party.

Central America has no large economies like Brazil's or Argentina's that could coalesce to counter US hegemony. Large-scale resistance to the FTAA and the likelihood that it will not meet US goals have led the US to push CAFTA. In December 2003, US trade negotiators herded four Central American countries — Nicaragua, Guatemala, Honduras, and El Salvador — into CAFTA. Costa Rica left the negotiations, refusing to open its service sector (domestic telecommunications, insurance, and tourism) to privatization bids by US-based corporations, but backed down under pressure and signed the pact a month later. CAFTA will abolish duties on 80 percent of US industrial and consumer goods, eliminate any remaining tariffs over the next 10 years,[73] and increase protections for intellectual property (trademarks, patents, and copyrights). Protections on sensitive agricultural products — such as US sugar and Central American white corn — will be phased out over 20 years.[74]

In the contentious 2004 election year, CAFTA faces an uphill battle in the US Congress. "Fast Track" legislation, renamed "Trade Promotion Authority" and passed in 2002, forbids amendments. Midwest sugar beet farmers and Southeastern sugarcane growers denounce CAFTA because it loosens import quotas that protect the domestic sugar industry.

CAFTA will affect Central America in ways similar to NAFTA's on Mexico. CAFTA negotiators promise development, decreased poverty, economic growth, and higher employment rates, the same promises made to Mexico during NAFTA negotiations. Unfortunately, the exact opposite has come to pass in Mexico, compounded by ecological disasters. Both agreements open the floodgates for GMOs to enter countries throughout Latin America, intensifying genetic contamination.

THE POLICY OF CONTAMINATION

What better way to control the world's food supply than by contaminating it with GMOs? If corporations can pollute the entire world's production of basic grains with DNA from GMOs, regulating them becomes an impossibility. If they can convince the world that contamination is widespread — even where it is not — then they can still argue that regulation is unnecessary. If implemented, CAFTA and the FTAA will inevitably spread genetic adulteration throughout the Western Hemisphere, providing excellent business opportunities for transnational corporations. They will have secured markets throughout Latin America and the Caribbean and member countries will be unable to block GE food imports from the US. They will no longer be able to produce their own untainted products and label them GE-free. This would hinder their ability to export their corn or processed products to Europe and other regions that restrict imports of GE foods. US transnational corporations would thus be able to systematically control the entire region's markets, determining to which countries communities can export their products.

In addition to manipulating trade policies, the US exploits international food aid agencies — particularly the US Agency for International Development (USAID) and the UN World Food Program — to force GE foods onto world markets, with manifold economic and environmental consequences (see Chapter 5). Dumping highly subsidized grains disrupts local markets, undermining the economic integrity of domestic farm economies. Many countries facing severe hunger are compelled to accept food aid; however, this "aid" often comes with strings attached, akin to the structural adjustment programs of the World Bank and International Monetary Fund (IMF). It often acts as a disincentive for farmers to cultivate their fields, building a stronger dependence on imports and giving US corporations a foot in the door.[75]

In Latin America, GMOs have been detected at content levels up to 90 percent in food aid sent to Nicaragua, Guatemala, Ecuador, Colombia, and Bolivia.[76] StarLink corn, approved only for cattle feed, showed up in food aid sent to Bolivia. More testing is needed to find out where other contamination has occurred.

CAFTA's Brother: Plan Puebla Panama

Plan Puebla Panama (PPP) is the infrastructural brother to the FTAA and CAFTA. Proposed by Mexican President Vicente Fox, the PPP is a 25-year "development" project that will stretch from the state of Puebla through Panama at a cost of at least $10 billion. Corporations write "free trade" agreements to create conditions favoring their investments and guaranteeing profits. They also demand the physical infrastructure needed to exploit the region's abundant resources, thus the PPP facilitates international trade and deepens corporate dominance in the region.

Under the PPP, railroads, dry canals, and five-lane highways would facilitate the flow of goods and services across borders. Also included is the privatization of electricity through a single regional energy grid, the proliferation of free trade zones, and increased militarization.[77] The Mexican government and the Inter-American Development Bank, the project's primary champions, promise that the PPP will bring development and prosperity to the region, but the proposed infrastructural "improvements" will not benefit those without vehicles or goods and services to sell. Rather, the PPP will aid mostly US-based corporations at the expense of Central Americans.[78]

A farmer from El Salvador explained, "These roads will only bring in the trucks that will run us over, they are not going to help the *campesino*."[79] Another *campesino* said, "The only advantage *campesinos* will see on this road is the increased presence of buses and trucks that we can sell chewing gum to."[80] New transport will open more of rural Mexico and Central America to industrial agriculture. Along with CAFTA and the FTAA, the PPP will help contaminate — thus destroying — indigenous and local varieties of corn and other crops throughout Central America.

RESISTING GENETIC CONTAMINATION AND FREE TRADE AGREEMENTS

A comprehensive plan to eradicate GMO contamination through-out the world needs to be implemented immediately. Communities, after being fully informed about GE dangers, must make the ultimate decisions about what to do. Refusing GE or untested seed, whether purchased or "donated" from the US, Canada, Argentina, or other GMO-friendly nations, should be the first line of defense.

A long-term strategy is needed to prevent further contamination, especially in the world's centers of crop origin and biodiversity, including Mexico and the rest of Central America. Ending the production of genetically engineered crops worldwide is a crucial step to safeguard the biodiversity of the crops and environments that form the shared heritage of the planet. Indigenous and *campesino* farming practices in Mexico and Central America need to be preserved and encouraged. Trade agreements and global financial institutions like the World Bank and the IMF must not be allowed to dictate what farmers produce or export to industrialized countries.

Agriculture is millennia older than free trade, corporations, and capitalism itself. Farmers do not need transnational agribusiness corporations dictating which seeds to sow or what kinds of crops to grow. Farmers do not need trade pacts to tell them which agricultural products they should export and import. Farmers must be free to choose the crops they grow and the seeds they cultivate. They are the protectors of crop biodiversity.

A deeper critique of genetic engineering, addressing economic hegemony and trade, is needed at the global level so that people can address the underlying causes of environmental devastation, including the contamination of crops through genetic engineering. The case of Mexico should serve as a warning to people everywhere about the consequences of "free trade" and crop contamination. NAFTA and the WTO should be repealed and future negotiations of "free trade" agreements cancelled. As the adverse effects of NAFTA multiply, people throughout the world are becoming more aware of the dangers that such pacts pose to national sovereignty, human rights, and the environment, indeed to the Earth's very survival. After a decade of NAFTA and the WTO, the world says, "*¡Ya Basta!* (Enough Already!)"

Biotechnology, Intellectual Property Rights, and the WTO

AZIZ CHOUDRY

"We are writing the constitution of a single global economy."
— Former WTO Director-General Renato Ruggiero[1]

O N JANUARY 1, 1995, THE WORLD TRADE ORGANIZATION (WTO) was established as the culmination of the Uruguay Round of trade negotiations of the General Agreement on Tariffs and Trade (GATT). A trading agreement between signatory countries, GATT was established in 1948 to set international rules regulating trade in goods among its members. By 2004, 148 governments had become WTO members.

With a secretariat in Geneva and a permanent staff of about 500, the WTO now incorporates GATT and over 20 other multilateral agreements, which all WTO members must sign, as well as several plurilateral agreements that only some members have signed. The multilateral agreements under the WTO are said to be a "single under-taking." This means that when a country signs one, it must accept all of them. The WTO administers these agreements, facilitates future trade negotiations, and enforces trade dispute resolution. A General Council acts as a dispute-settlement and trade policy review body. Other councils deal with different issues, operating through numerous committees. In mid-2003, one of the best-known WTO disputes to date was launched by the US against the European Union (EU) over its *de facto* moratorium on approvals for genetically engineered (GE) imports and crops.

The core principles of the WTO are "national treatment" and "most favored nation" (MFN) trading status. National treatment means that foreign products and services should receive no less favorable, and possibly better, treatment than domestic products and service suppliers, making it impossible for members to protect local markets. Under MFN, a signatory country must give all WTO members the best treatment it has granted to any one of them.

During the Uruguay Round (1986-1994), industrialized countries, spurred on by transnational corporations, aimed to make GATT the most powerful instrument controlling trade, while incorporating many new areas not previously treated as trade issues. The Reagan administration blamed its economic problems on protectionism and unfair competition, especially from the EU, Japan, and newly industrialized countries like South Korea and Taiwan. It insisted that the scope of trade talks cover new areas of interest to US-based transnationals. Thus, intellectual property rights (*e.g.*, patents, copyrights, trademarks), services (*e.g.*, water, utilities, health care), agriculture, and aspects of investment would fall under WTO jurisdiction when they were designated "trade-related."

"Developed" countries, which were losing out to Brazil, South Korea, and other Southeast Asian countries in traditional trade in goods, were determined to make up for their losses by capturing markets for their service industries and investments. Thus, government policies setting national social and economic priorities were often redefined as barriers to trade, and targeted for dismantling through the negotiation process. Northern governments promised their counterparts in the global South better market access for Southern agricultural, garment, and textile products in return for major concessions in other areas, including the acceptance of WTO agreements on intellectual property rights, services, and trade-related investment measures. Such horse-trading served to open new markets for largely Northern-based corporations across the world, while the promised market access for Southern countries' exports remained largely a mirage.

Neoliberalism describes both the ideology and strategy behind the policies promoted by the WTO and other vehicles for "globalization." It advocates total freedom of movement for capital, goods, and services, sees everything and everyone as a tradable commodity, and argues that market forces must be left to rule, free from interference by government or communities. That includes the food we eat, the water we drink, and life itself. Social and environmental issues only make it onto the WTO agenda if defined in "market friendly" terms. WTO agreements oblige governments to make serious legislative and regulatory reforms that impact a wide range of domestic policies.

New countries seeking to join the WTO have to go through a painful process of structural adjustment and so-called liberalization. Typically, structural adjustment programs (SAPs) imposed by the World Bank, International Monetary Fund, and other international financial institutions include privatizing state-owned enterprises and services, slashing public spending, orienting economies towards export, eliminating "barriers" to trade and investment, increasing interest rates and taxes, and slashing subsidies on basic consumer items like food, medicine, and fuel. While this model has worked spectacularly well for global corporations and foreign investors, it has been an abject failure for most of the world's peoples. Trade and investment "liberalization" are SAP conditions, pushing many countries to join the WTO as a way of demonstrating to creditors their commitment to a market economy. But to join, countries are often forced to sign radical bilateral trade and investment agreements with WTO countries, like the one Vietnam signed with the US in 2001. Besides opening Vietnam's economy to US agricultural and industrial exports, this agreement gave US firms wide access to Vietnam's markets for financial, telecommunications, distribution, audio-visual, legal, accounting, engineering, computer, market research, construction, educational, health, and tourism services.

If trade disputes cannot be settled by mediation and consultation, the WTO system lets countries challenge each others' laws as violations of WTO rules. Cases are decided in secret by a panel of three trade bureaucrats. These are usually former GATT or other high-level trade officials. Hearings are secret, and documents and other evidence presented to dispute panels are confidential unless parties decide to release them. Only WTO member governments have legal standing; trade unions, indigenous peoples, and non-governmental organizations (NGOs) have no right to participate.

Every environmental or public health law challenged at the WTO Dispute Settlement Body has been ruled illegal. After a ruling, assuming that any further appeal to the WTO Appellate Body is unsuccessful, the losing country can either change the law to conform to WTO requirements, pay compensation to the winning country, or face non-negotiated trade sanctions. The winning country can apply trade sanctions against the non-complying country in any area covered by the WTO Agreement. Only a unanimous vote of all member governments can overturn a ruling. The US has been the world's most frequent user of the WTO dispute settlement system.

The WTO is supposed to operate by consensus, but it is dominated by powerful governments (the US, EU, Japan, and Canada, known collectively as the "Quad"), which try to impose their decisions on other members. In contrast, smaller and poorer countries, which are often excluded from key sessions, have limited access to negotiations. They cannot afford to maintain the teams of negotiators and trade lawyers necessary to make their voices heard on trade policies. Dispute resolution processes, too, are slanted against such countries, which lack the resources to defend themselves against complaints by rich countries. Often, just the threat of a complaint forces them to settle a dispute in favor of transnational enterprises, and against the interests of the majority of their citizens.

Negotiating positions and the details of individual government commitments at the WTO are closely guarded secrets. World Bank economist Bernard Hoekman and co-author Michel Kostecki write that WTO negotiations are meant to push governments further than they would otherwise go, and to counteract pressures to backtrack.[2] This is much easier when they can make commitments and sign agreements without the interference of domestic pressure groups.

TRIPs: Intellectual Property and Bio-Colonialism

The Agreement on Trade-Related Aspects of Intellectual Property Rights, known as TRIPs, really has nothing to do with "free trade." It is a protectionist tool that requires all WTO members to guarantee the protection of patents for at least 20 years. It was packaged as an anti-counterfeiting proposal for companies that wanted to stop fake brand name clothing, music, and videos, but it prohibits measures commonly used to facilitate technology transfer such as compulsory licensing. This is when a government gives a manufacturer a license to produce something for which another company holds a patent or exclusive rights, in return for the payment of a royalty. Its goal is to introduce generic competition and to drive prices down. TRIPs also covers copyrights and related performance rights, layouts of integrated circuits, "geographical origin" indicators (as for wines and gourmet cheeses), trademarks, and industrial designs. It sets the stage for broadening patent rights for genetically modified organisms (GMOs) and other products of biotechnology.

The concept of intellectual property rights itself has been strongly criticized, especially by indigenous peoples. It is based in a Western scientific paradigm of reductionism — the tendency to reduce all phenomena to their component parts. As Maori researcher Aroha Te Pareake Mead notes, "[E]ach level of reduction presents an increased scientific opportunity."[3] She explains that intellectual property laws "do not regard existent indigenous knowledge as being an intellectual property and deserving of protection, rather they consider such knowledge as 'common' and define human intervention based on what non-indigenous peoples 'add' to what has existed for generations."

The notion of intellectual property rights emerged from interlocking Western doctrines of commerce, science, and the law, which were used to justify and expand colonization. The idea that knowledge can be created, owned, or sold by a single inventor conflicts with many indigenous and other non-Western views that knowledge is inextricably linked to culture, spirituality, identity, and place, and is created communally over time. Intellectual property rights commodify and privatize knowledge for exclusive exploitation and private profit.

The Intellectual Property Committee (a coalition of 13 large US corporations, including DuPont, Pfizer, IBM, General Motors, Rockwell, Bristol-Myers, and Merck) worked with US trade representatives on a proposal to standardize global intellectual property laws along US lines, and make them enforceable under the WTO. This followed failed attempts during the 1980s to negotiate tighter rules and a global patent regime on intellectual property at the Geneva-based World Intellectual Property Organization (WIPO). There, Southern governments had claimed that they needed the same rights to access knowledge and technology that benefited richer, industrialized countries when intellectual property laws were weak. The TRIPs agreement ultimately discarded such claims.

Ninety-six of the 111 members of the US delegation negotiating on intellectual property during the Uruguay Round came from the private sector. Diplomats in Geneva say that the pharmaceutical industry drafted much of TRIPs, while the US government was its lead advocate. At the start of the Uruguay Round, the US negotiator appointed to head the delegation on what was to become the WTO Agreement on Agriculture was Dan Amstutz, former vice-president of agribusiness giant Cargill, who recently headed the USAID-driven "reconstruction" of Iraq's agriculture.

Robert Shapiro was chair of Monsanto while also leading the President's Advisory Committee for Trade Policy and Negotiations. Mickey Kantor, US trade representative for much of the Uruguay Round, subsequently became a Monsanto board member. Marcia Hale, a former assistant to President Clinton and director for intergovernmental affairs, was director of international government affairs for

Monsanto. Clayton K. Yeutter, a former secretary of agriculture and US trade representative, who led the US team in negotiating NAFTA and helped launch the GATT Uruguay Round, joined the board of directors of Mycogen Corporation. Mycogen's majority owner is Dow AgroSciences, a wholly owned subsidiary of the Dow Chemical Company.[4] The web of interconnections between industry and successive US administrations' trade negotiators ensures that private (often monopoly) interests will trump those of people and the planet.

TRIPs strengthens the hand of private companies in claiming monopoly rights and securing huge benefits from biopiracy. Private sector researchers, agribusinesses, and pharmaceutical corporations are appropriating indigenous communities' heritage, while those who developed and nurtured them receive no benefits. TRIPs forces all WTO member countries to comply with a minimum standard set of laws protecting the technological monopoly of transnational corporations (mostly from industrialized countries), which own most patents, including patents on seeds and genetic sequences. TRIPs goes hand in hand with WTO commitments to "liberalize" agricultural trade, further expanding agribusiness control over food systems and biodiversity. It was the first international instrument to require intellectual property rights protection over life-forms.

US case law set the international precedent for patenting genetic material. In 1980, the US Supreme Court allowed the patenting of microorganisms. In 1985, life patenting was extended to include plants. In 1987, the US Patent Office ruled that all animals, including human embryos and fetuses, were also patentable. TRIPs and similar provisions in other regional and bilateral agreements are being used to extend and implement US-style patent law worldwide. According to the US, countries failing to adopt such laws are engaged in "unfair trading practice" using "non-tariff trade barriers," and deserve trade sanctions. Early TRIPs targets were Japan and newly industrializing countries in East Asia that had copied US technology, especially in the automotive and computer industries. But the ongoing plunder of the South's biodiversity, without any compensation or benefits for the communities from which the "raw material" originated, is not considered unfair.

Before the Uruguay Round, most nations chose not to recognize patents on food, pharmaceuticals, or other products considered basic human needs. The policies of many governments were shaped by specific ethical and socio-economic considerations. TRIPS requires governments to allow microorganisms and microbiological processes (as well as non-biological processes) to be patented. It requires governments to ensure that plant varieties be protected by patents or a *sui generis* system (*i.e.*, some other form of plant variety protection) or a combination of the two. Many governments have had to enact domestic plant variety protection laws based on the model of UPOV (International Union for the Protection of New Varieties of Plants). While a country may exclude from patentability plants, animals, and "essential" biological processes for the production of plants and animals, the outcome of a pending WTO review of TRIPS could change that. The integrity of life is now deemed to be a mere barrier to trade and scientific progress. "Developed" WTO member countries were required to apply TRIPs by January 1, 1995. "Developing" member countries had until January 1, 2000, while "least developed" country members have a January 2006 deadline and are able to apply for extensions from the WTO Council on TRIPs.

Current intellectual property systems reward individual "inventors" for products, processes, or innovations relating to genetic material derived from plants, animals, or organisms — but not communal knowledge, such as that shared and handed down in indigenous communities. When genetic material is processed in corporate labs it is named, called an "invention," and usually patented, bestowing exclusive marketing rights on its "owner." Broad patents are being granted for plant varieties, covering ownership of "traits" and "characteristics." Seed companies stand to benefit greatly from this monopoly, while innovations in the use of such plants or trees by small farmers' and indigenous peoples remain unrewarded. Increasingly, they will be prohibited from using and saving their own seed and forced to buy them from companies like Monsanto.

Genetic Resources Action International (GRAIN) points out that WIPO has played a key role in implementing TRIPs standards in the South. This has included drafting and recommending "TRIPs-plus" legislation. West African countries were advised to implement TRIPs well ahead of their extended compliance schedule as Least Developed Countries (LCDs), and advised against using the rather limited flexibility TRIPs allows in compulsory licensing or parallel imports.[5] For example, WIPO did not inform Cambodia's government that, as an LDC, it was not obliged to grant patents on pharmaceuticals before

2016. GRAIN warns that WIPO is another arena for the fight over intellectual property rights; a new international patent treaty, the Substantive Patent Law Treaty, is currently being negotiated.[6]

Meanwhile, consolidation of the biotech industry continues, creating mega-corporations with global tentacles. In 1996, Robert Fraley, then president of Monsanto's Ceregen Division, explained to the US magazine *Farm Journal* the company's strategy of taking over scores of plant-breeding institutes and smaller biotechnology firms. "What you are seeing," he boasted, "is not just a consolidation of seed companies, it's really a consolidation of the entire food chain."[7]

Transnational corporations are appropriating the knowledge of local communities, especially in the tropics and subtropics of the Americas, Asia, and Africa. The peoples of these regions have developed, protected, and nurtured biodiversity for many centuries, and they are the source of many of the world's food crops and medicines. For years, pharmaceutical companies have been seeking access to indigenous communities' knowledge to find plants and traditional ways of using them. This offers a far better chance of finding potential pharmaceutical products than random screening. University of Illinois at Chicago scientist Norman Farnsworth says "there are 121 prescription drugs in use today, which come from only 90 plant species. About 74 percent came from following up folklore claims. There are 250,000 species of plants in the world, so there are clearly many more jackpots to be found."[8]

Owning a life form patent has a far greater reach than owning an individual sheep or tree. US researchers Hope Shand and Martin Teitel note that the distinction "can be likened to the difference between owning a lake and owning the chemical formula for water. A patent holder for water's chemical formula would have the legal right not only to decide who could have access to a particular lake, but to water anywhere, and to the use of the chemical formula for any purpose."[9]

Indigenous peoples are at the forefront of challenging bio-colonialism. The president of the Guaymi General Congress in Panama has said, "I never imagined people would patent plants and animals. It's fundamentally immoral, contrary to the Guaymi view of nature, and our place in it."[10] Similarly, local communities from Southern Mexico through Central America are resisting the Meso-American Biological Corridor project, billed as an ecological protection plan.

Running alongside the proposed Plan Puebla Panama (see Chapter 1), it is backed by the World Bank as a "green" front to open up the region's rich biodiversity to transnational corporations seeking profits and monopoly control. Indigenous peoples living within the corridor's boundaries face dislocation and the likelihood that companies will steal the microorganisms and plants from the forest, even as they open up the region to more ecologically destructive forms of development.

TRIPs, the proposed intellectual property chapter of the FTAA — and similar provisions in bilateral and regional trade agreements, which often contain "TRIPs-plus" provisions that go even further — are all tools to expand, intensify, and lock in a regime of monopoly control over life itself. And because intellectual property rights are often included in the definition of "investment" in bilateral investment agreements, any perceived failure to comply with corporate demands for patent protection on genetic material in a signatory country could lead to a NAFTA-style investor-state dispute. Under NAFTA's investment rules, companies can bring a claim against any government (in effect, a corporate lawsuit) before a special dispute settlement body. A proposed WTO investment agreement, promoted by the EU, contains a similar definition of "investment" and would have the same effect.

Frustrated in attempting to develop their own technologies, Southern countries simultaneously pay massive royalties to Northern corporations, often for products derived from ideas and biodiversity originating in the South. The knowledge created and shared by indigenous and traditional communities, and nature itself, is being commodified and privatized at a frightening pace.

With WTO negotiations thus far failing to deliver as much as many businesses want, the US and other governments are increasingly turning to bilateral trade and investment agreements. These have the advantage of slipping underneath the radar of many NGOs and popular movements that mobilized against the WTO and the FTAA. Prior to the WTO ministerial conference in Cancun in 2003, where global trade talks collapsed without agreement, the US had concluded comprehensive bilateral free trade and investment agreements with Chile and Singapore. These came into force on January 1, 2004. The US has either concluded, is negotiating, or plans to negotiate other free trade and investment agreements with a number of other countries.

These bilateral trade agreements contain broad "NAFTA-plus" definitions of investment, which throw the door wide open for disgruntled investors to take a case to a dispute tribunal. Intellectual property provisions in these agreements go even further than TRIPs. They severely limit the grounds for allowing use of compulsory licensing of medicines, and effectively extend 20-year drug patent monopolies by an additional five years, threatening access to affordable medicines, including HIV/AIDS drugs. Moreover, this "TRIPs-plus" approach does not allow for plants and animals to be excluded from the patent laws of signatory countries. While TRIPs sets a minimum standard for intellectual property protection, these other agreements are imposing an industry-driven agenda through the backdoor, locking countries into even more stringent intellectual property standards.

The section of the TRIPs agreement that explicitly extended patents to cover living organisms, Article 27.3(b), was supposed to be reviewed in 1999. A review of the entire agreement was also supposed to take place, but neither review has occurred. A 1999 statement from a group of African countries said developing countries should be given more time to consider the implications of implementing TRIPs. They called for new wording in TRIPs that recognizes the right of countries to "satisfy their need to protect the knowledge and innovations in farming, agriculture and health and medical care of indigenous people and local communities." It also urged inclusion of farmers' rights to save and exchange seed and sell their harvests in a revised TRIPs agreement, and highlighted a serious inconsistency in TRIPs: Although plants and animals could be excluded from patenting, microorganisms could not.

In June 2003, the African group made another formal submission on TRIPs. It sought to outlaw the patenting of all life forms, including microorganisms, plants, and animals. It urges an end to the patentability of non-biological and microbiological processes for animal and plant production. Their view is based on a desire to "protect important policy goals relating to food security, nutrition, the elimination of rural poverty, the integrity of local communities... the protection of genetic resources, traditional knowledge... [and] cultural rights."[11] People around the world continue to devise mechanisms and tools to safeguard traditional knowledge and control genetic resource use and development.

TRIPs vs. Public Health

When governments have sought to respond to health crises, especially HIV/AIDS, by permitting the use and import of affordable generic drugs, they have faced litigation both by pharmaceutical companies and the WTO. The powerful US drug cartel, the Pharmaceutical Research and Manufacturers of America (PhRMA), lobbied to ensure that the US administration would carry out its agenda. This led to threats of trade sanctions against countries — including India, South Africa, Brazil, Argentina, and the Dominican Republic — over compulsory licensing or parallel importation laws, as well as a notorious court case against South Africa. Aggrieved by South African legislation which allowed the country to import and manufacture cheaper anti-AIDS drugs, PhRMA sued the South African government on behalf of US drug companies. Under mounting pressure, within South Africa and internationally, the case was suspended. In June 2001, an EU and US WTO complaint against Brazil's violation of drug patents — in its effective and internationally-praised anti-AIDS program — was withdrawn after worldwide pressure.

A 2001 WTO ministerial declaration on TRIPs and public health said that TRIPs should be interpreted in a way that enables governments to protect public health and "promote" access to medicines for all.[12] The declaration was non-binding, and of uncertain legal value, but was used as a bargaining chip to get governments of Southern countries, many resisting new and expanded world trade talks, to acquiesce to the US-EU push for a another WTO round. In 2002, the US, acting on behalf of its pharmaceutical corporations, rejected a proposed TRIPs and public health agreement at the WTO that would have allowed developing countries to issue compulsory licenses for the manufacture and importation of generic drugs to combat a number of health problems.

Just prior to the WTO meeting in Cancun, however, a "deal" on TRIPs and public health was supposedly struck. Promoted by some free trade cheerleaders as an "historic breakthrough" and proof that the WTO "has a heart," the deal was a morale booster for the institution, which had missed several negotiating deadlines in the run-up to Cancun. But it was mainly a political victory for the WTO, not for the millions of people denied access to lifesaving drugs through unjust patent regimes.

With US backing, pharmaceutical corporations played hardball, making sure that the accord contains so many conditions that many critics believe it will continue to prevent drugs from reaching those who need them. "The ability to import cheap generic drugs is being undermined by adding more conditions and restrictions so there will probably be only very limited use of this agreement," said Bernhard Herold of the Berne Declaration, a Geneva-based organization working for equity in North-South relations.[13]

The deal is "designed to offer comfort to the US and the Western pharmaceutical industry," says Ellen 't Hoen of Medicins Sans Frontieres (Doctors Without Borders). "Unfortunately it offers little comfort for poor patients. Global patent rules will continue to drive up the price of medicines."[14] Yet, the perception that there was a significant concession on the part of the US and other Northern pharmaceutical-manufacturing nations is being used as leverage to force more commitments from Southern delegations.

AGREEMENT ON AGRICULTURE

Another WTO agreement with implications for the spread of GMOs is the Agreement on Agriculture (AOA). This was heralded as a means to provide greater access to world markets for agricultural products for all countries by reducing tariffs and other trade barriers, as well as farming subsidies. It aims to "establish a market-oriented agricultural trading system ... [and] reductions in agricultural support and protection ... resulting in correcting and preventing restrictions and distortions in world agricultural markets."[15]

The removal of tariffs and other import restrictions has opened up countries to floods of agricultural imports, including GMOs. In June 2000, a group of 11 developing countries told a WTO special session of the Committee on Agriculture that the trade liberalization triggered by the Uruguay Round had broken the agricultural backbone of many countries, undermining food security, peoples' health, and sovereignty.[16] The AOA also requires uniformity in sanitary and phytosanitary (plant health) measures that cover food hygiene and inspection (see below). Starting in 1995, these commitments were supposed to be implemented over a six-year period (1995-2000) by "developed" countries or 10 years (1995-2004) by "developing" countries.

After a decade, Mexican *campesinos'* (small farmers) experience of NAFTA leaves them with no illusions about the WTO's promises concerning agricultural "free" trade. A flood of cheap, subsidized US corn has entered their market priced below the production cost. *Campesinos* cannot compete, leading to massive displacement, poverty, and hunger, pushing people into cities and *maquiladoras* (sweatshop factories), and forcing many to risk their lives crossing the increasingly militarized US border in search of work. Moreover, an estimated 30-40 percent of this corn is genetically modified, and there has already been serious contamination of indigenous corn varieties from GMO corn imports in Oaxaca, as discussed in Chapter 1 of this book.

A global peasant and small farmers' movement, La Via Campesina, has mobilized for food sovereignty and against biotechnology, the AOA, and corporate control of agriculture. Each country, it argues, should have the right to define its own agricultural policies to meet domestic needs. This should include the right to prohibit imports to protect domestic production, and genuine agrarian reform to provide peasants and small to medium-sized producers with access to land.

Via Campesina calls for the WTO to get out of agriculture, urging the removal of rules on food production and marketing from WTO negotiations and other regional and bilateral agreements. It states: "A profound reform of the WTO in order to make it respond to the rights and needs of people would mean the abolition of the WTO itself! We do not believe that the WTO will allow such a profound reform."[17]

HUNGRY FOR PROFIT: THE US COMPLAINT ON GMOS

Through its "food aid" policies and uses of the WTO and other agreements, the US — backed by its biotech/agribusiness corporations — wants to force genetically modified seed, grain, and foodstuffs into all the world's markets, fields, and stomachs by deliberate genetic contamination, and by targeting countries which have taken principled stands against genetic engineering.

In August 2001, pressure from the WTO, the US, other Northern governments, and some domestic businesses forced Sri Lanka to indefinitely postpone a proposed ban on GE food imports. Sri Lanka wanted more time to assess associated health risks.[18]

In May 2003, the US initiated a dispute under the WTO, challenging the EU's *de facto* moratorium on approvals for GE imports and crops, in place since 1998. In addition to citing the GATT and AOA, the US claimed that the moratorium breached the WTO Agreement on Sanitary and Phytosanitary Measures (SPS). US Trade Representative Robert Zoellick blamed the EU for the refusal of US food aid by African countries, and asserted that the US is promoting GE food to help feed the world. In fact, human misery has become a marketing tool for US agribusiness.[19]

After consultation failed to resolve the matter, a dispute panel was established in late August 2003. The US was joined by Canada, Argentina, Peru, Colombia, Mexico, New Zealand, Australia, India, Brazil, and Chile. Egypt initially supported the complaint, but pulled out, only to find itself the target of US displeasure. Egypt's involvement in the complaint was important for the US, since the Bush administration claimed that Africa would suffer as a result of the European GMO moratorium. However, the Egyptian Ministry of Foreign Affairs was also concerned about relations with the EU. After warmly praising Egypt and promising that it topped the list of Middle Eastern countries for the US to negotiate a free trade agreement with, Cairo was told that the US would now not pursue trade negotiations.[20]

US Agriculture Secretary Anne Veneman, a former director of the biotechnology company Calgene and member emeritus of the International Food and Agricultural Trade Policy Council (a policy group financed by Monsanto, Cargill, ADM Kraft, and Nestle[21]), was more candid than Zoellick about the dispute's real causes. "With this case, we are fighting for the interests of American agriculture," she remarked.[22] Of course, she actually meant US corporate agribusiness.

The SPS agreement recognizes that countries can regulate crops and food products to protect health and the environment, and it allows domestic measures to be higher than international standards. Invoking a need for "sufficient scientific evidence," however, the agreement is being used to restrict countries' ability to independently set such measures and approval procedures. The US argues that EU actions are an unjustified regulation to thwart trade in "safe, wholesome, and nutritious products," in other words, untested GE food. Meanwhile, some Southern food exporting countries accuse the North of using differing SPS norms as a new way to block their products, even as they claim to be improving agricultural market access.[23]

In an earlier case, the US and Canada launched a WTO dispute against EU legislation that banned the use of certain growth-producing hormones in beef cattle. In January 1998, the WTO Appellate Body issued a report stating that the ban was not based on a risk assessment, as required by the SPS. This dispute set a precedent for using the WTO to attack all kinds of domestic regulations based on health concerns, which are viewed as barriers to trade. Since July 1999, the US and Canada have imposed trade sanctions against the EU, consisting of an increase in tariffs for a selected list of products in the amount of US $116.8 million and $11.3 million Canadian respectively.[24]

In complaints brought under SPS — a business-oriented agreement aimed at deregulation — the WTO rulings could compel a nation to choose between lowering its health standards for humans, animals, or plants; compensating another government whose exports are limited or blocked by the stricter standard; or permitting that country to impose additional trade restrictions on exports from the nation with the higher standard. Along with the WTO Technical Barriers to Trade (TBT) agreement (which covers technical regulations, product standards, and testing and certification procedures), SPS opens the way for attacks on national measures that address consumer concerns, such as labeling products containing GMOs. Pressure for downward harmonization (see below) is built into the agreement.

TBT is supposed to ensure that standards and procedures do not create "unnecessary obstacles" to trade. It commits WTO members to using appropriate international standards, largely set by industry, in their technical regulations. Under this agreement, governments must notify the WTO Secretariat of any proposed new measures, including information on the objectives and rationales behind the measures and on the products covered. This opens them for comment and amendment by other WTO member governments.

Backers of neoliberal globalization are eager to "harmonize" food and product standards, replacing diverse national standards with universal ones shaped by industry. High standards are likely to be replaced by the lowest common denominator. The WTO and other trade agreements currently base their food safety standards on those set by Codex Alimentarius,[25] a commission jointly administered by the UN's Food and Agriculture Organization and the World Health Organization.

Codex plays a major role in defining SPS and TBT standards for food safety and quality, and making them more uniform at the international level. For example, it allows 50 times more DDT to be used or left in residual amounts on peaches and bananas, and 33 times more DDT to be applied on broccoli than US Environmental Protection Agency (EPA) standards. It is the primary reference for the WTO in SPS measures. Northern government delegations to Codex have included corporate representatives from Nestle, Coca-Cola, Pepsi, Cargill, and SmithKlineBeecham. The 11th session of the Codex

Committee on Fresh Fruits and Vegetables met in Mexico City in September 2003 while the WTO Ministerial began in Cancun.

A current priority for the biotechnology committee of the powerful US Council for International Business is to "[s]upport Food Industry Codex Coalition activities, especially those related to precaution, risk assessment, and labeling [and to d]evelop a long-range plan of action for ensuring the continued use of modern biotechnology in various applications, including non-agricultural biotechnology."[26] On July 9, 2003, Codex adopted "Principles and Guidelines on Foods Derived from Biotechnology." These lay out general principles intended to make the analysis and management of risks related to foods derived from biotechnology uniform across Codex's 169 member countries. The guidelines concern food safety, and do not address environmental risks at all.

Big business and the Bush administration are irked at the use of the "Precautionary Principle," which provides a more cautious basis for some governments' positions and policies, including the long-standing EU moratorium on approving new GE crop varieties. This principle argues that precautionary measures should be taken when an activity raises a threat of harm to human health or the environment, even if some cause and effect relationships are not yet scientifically established. It also places the burden of proof on the proponent of the activity to prove its lack of harmful effect.[27]

This principle is under sustained attack from both the US and WTO dispute panels. The US and other governments that have joined the complaint against Europe have a warning for the world: Don't even think about imposing restrictions on GE foods or we will come after you, big time. So much for people having the right to know and decide what they eat and grow. So much for principles of any kind.

The World Bank:
Biotechnology and the "Next Green Revolution"

BRIAN TOKAR

HE WORLD BANK, WHICH CELEBRATES ITS SIXTIETH ANNIVERSARY in 2004, has underwritten several of the most environmentally devastating projects ever undertaken in the name of progress and economic "development." Wherever people are displaced and communities are undermined in the name of development, World Bank funds are almost always part of the mix. The Bank has supported deforestation, hydroelectric projects, and oil drilling in the Amazon rainforest; huge dams and oil and gas pipeline construction in Africa; and massive water diversion schemes such as India's notorious Narmada Valley dam complex.[1]

Bank-funded projects have uprooted millions of people from their traditional lands, irrevocably destroyed fragile and unique ecosystems, and created footholds for transnational corporations throughout the so-called "developing world." In the past decade, Bank officials have adopted the language of "sustainable development" and withdrawn funds from a few of the most notorious projects in their portfolio. But few advocates for the environment or the rights of indigenous peoples have any illusions that the institution's priorities ever substantively shifted along with its rhetoric.

The idea of "sustainable development" emerged from policy discussions at the United Nations (UN) throughout the 1980s, and was popularized in the lead-up to the UN's high-profile environment and development summit in Rio de Janeiro in 1992.[2] Merging the language of long-term sustainability from the environmental movement with the "development" discourse of neo-colonialism, sustainable development became a rationale for advocating the continued expansion of capitalist market economies in the global South, while paying lip service to the needs of the environment and the poor.

The popular notion of environmental sustainability as an alternative to limitless economic growth was transformed into a rhetorical justification for economic growth in defiance of environmental and social limits. Almost imperceptibly, the discussion shifted from how to stem environmental destruction to finding new ways to sustain economic growth.[3] Former US Defense Secretary and Vietnam War planner Robert McNamara introduced a focus on "poverty reduction" during his tenure as Bank president (1968-1981); since the mid-1990s, virtually every activity of the Bank, however controversial, has been justified as aiding both "poverty reduction" and "sustainable development."

In the agricultural sector, the Bank's strategy has helped displace people engaged in subsistence and local market-oriented production, in favor of commercial production for global markets. In virtually every corner of the so-called "developing world," the Bank has underwritten policies that diverted once-independent farmers toward the chemical-intensive production of cash crops. Bank officials say their goal is to "broaden farmers' rights and opportunities, and to help them create livelihoods of their own choice," beyond the "single option" of subsistence.[4]

For marginalized farmers throughout the world, this has brought an increasing dependence on unstable world crop prices, rising indebtedness for costly equipment and chemicals and, often, the forced removal of people from traditional lands that have sustained their communities for countless generations. Rather than helping alleviate poverty, this kind of "development," in the words of Japanese economist political analyst Ichiyo Muto, "has so far only transformed undeveloped poverty into developed poverty, traditional poverty into modernized poverty designed to function smoothly in the world economic system."[5]

The Bank's policy of shifting developing countries toward cash crop production began in the 1950s, with an initial focus on specialty crops such as cocoa, rubber, and palm oil. With the development of the so-called "high yielding varieties" of wheat and rice in the 1960s, Bank lending was often limited to governments that promoted the use of the new seeds and accepted their dependence on mechanization and costly chemical inputs. Seeds were often given away and enterprising farmers were offered attractive loan packages subsidized by the Bank.[6] These policies came to define the so-called "Green Revolution" of the 1960s and Seventies.

While these new crops brought significant short-term increases in agricultural productivity, their use lowered water tables and severely threatened crop diversity. The dwarf characteristics of these non-indigenous crop varieties deprived farmers of important agricultural byproducts, including sufficient fodder to feed livestock. The virtually endless need for new equipment and other inputs buried farm families in unsustainable long-term indebtedness.[7]

World Bank lending for agricultural projects declined from 30 percent of the Bank's portfolio in the 1980s to only 10 percent during the Nineties, but still amounts to nearly $3 billion per year, the largest source of agricultural development funds in the world.[8] Close to half of the Bank's portfolio is in structural adjustment loans tied to specific changes in the host country's economic policy; these are essentially identical in scope to the Structural Adjustment Programs (SAPs) imposed by the Bank's sister institution, the International Monetary Fund (IMF). Structural adjustment compels countries to reorient their economies toward the repayment of international debts, including IMF and World Bank loans, usually at the expense of public services, environmental protection, and local production for local needs.

In the case of agricultural lending, the Bank mandates debtor nations to shift agricultural production toward cash crops for export, liberalize agricultural trade, and remove public subsidies for staple food production, replacing fixed prices for staple goods with market-determined ones.[9] Producers are shifted from subsistence food crops toward fruits, vegetables, and flowers grown for the world market. Under SAPs, credit is offered to individual producers and denied to traditional communal activities, destabilizing rural societies and encouraging unprecedented concentrations of individual land ownership. Loans are often tied to specific production methods, including the increased use of hazardous pesticides and other costly materials.

The Bank's current support for biotechnology in agriculture has two aspects. The first encompasses technical assistance and "capacity building" for governments, aimed at facilitating the introduction of new biotechnologies and establishing biotech-friendly regulatory regimes. The second is direct support for biotech research. The Bank's capacity building agenda has five main areas of activity:

- Evaluating the potential of biotechnology to address local problems, especially through the use of cost/benefit and risk analysis;
- Promoting partnerships between corporations, private funding sources, and public agencies;
- Designing and implementing regulatory systems;

- "Educating" farmers and consumers; and,
- Promoting international cooperation in regulatory policy, financing, and technology transfer.[10]

The Bank also offers technical training, policy and management advice, analysis of countries' regulatory systems, and consultations with representatives from various social sectors to discuss proposed policies and their likely impacts. As we will see, the underlying assumption is often that biotechnology is the wave of the future, and that rational public policy can only serve to facilitate its development.

The Bank has provided some $2.3 billion in direct loans for research, of which $50 million is for biotechnology.[11] This includes some less controversial techniques, such as tissue culture and the use of DNA markers to assist plant breeders, but also genome mapping and transgenics. Less controversial pursuits are often viewed as "stepping stones" to more "advanced" applications, such as the development of new genetically modified crops. Over 80 percent of research funding is committed to six key countries: India, Kenya, Brazil, Indonesia, Peru, and Ethiopia. Some $20 million — 40 percent of the total — is for projects in India, where the Bank supports development of insecticide-producing Bt rice, as well as genetically engineered (GE) varieties of cotton, pigeon peas, chickpeas, and various horticultural crops.[12]

Also in India, the Bank has provided assistance for a controversial project in which genes from amaranth seeds have been spliced into the DNA of potatoes to increase the potatoes' protein content. This project was announced with great fanfare in 2003. While the protein content of the potatoes reportedly increased by nearly half, they contained a small fraction of the amount found in whole amaranth, or even wheat and rice.[13] The peas, lentils, and other legumes that are an important part of traditional Indian diets — but have been marginalized in cash crop-oriented agricultural development projects — provide even more protein. As with the $100 million effort to develop a GE "golden" rice containing increased beta carotene (a Vitamin A precursor), biotechnologists are promising a high-tech "cure" to hunger, while ignoring far more realistic and readily available solutions.

In Kenya, the Bank has provided support for a project largely financed by the US Agency for International Development (USAID), along with Monsanto and a number of other private donors. The stated aim is to offer GE varieties of sweet potatoes, a staple crop in rural areas that rarely attracts the interest of corporate researchers. After 11 years of research, which created a very high public profile for the Monsanto and USAID sponsored Kenyan researcher Dr. Florence Wambugu, only one local sweet potato variety has been genetically modified, imparting resistance to a virus that farmers routinely fend off by far less invasive means.[14] Under field conditions, this potato failed to demonstrate any significant virus resistance.[15] Indeed, more than 20 years of Bank agricultural projects in Kenya have failed to meaningfully assist poor farmers, according to the Bank's own analysts.[16] In most other countries, Bank-funded biotechnology research is largely directed toward technological capacity building, genetic analysis, and research support for the regulatory sector.

PROMOTING THE "NEXT GREEN REVOLUTION"

The World Bank's interest in biotechnology emerged in the 1980s, just as corporations such as Monsanto were beginning to shift their research priorities toward developing new transgenic crops. It began funding agricultural projects with distinct biotechnology components in 1982, and commissioned a study in 1988 to "assess the contribution that biotechnology might make to agricultural productivity, and to identify the socioeconomic, policy and management issues that might impede its successful application."[17] The study culminated in an international seminar in Canberra, Australia, in 1989, as well as a report published in 1991. The advisory committee that reviewed the final report included Val Giddings, now vice president of the Biotechnology Industry Organization, and Gabrielle Persley, who has continued to advise the Bank on biotech policy, while serving on the board of the industry-sponsored International Service for the Acquisition of Agribiotech Applications (ISAAA; see below), and also as the executive director of the AusBiotech Alliance in her native Australia.[18]

In many ways, this report defined the Bank's biotechnology agenda for the next decade and beyond. The focus then, as now, was on facilitating the adoption of biotech methods in the so-called "developing world," with an emphasis on tissue culture and advanced diagnostics, but also on genetic engineering and gene mapping. The report predicted that biotechnology would bring "modest but continued increases in productivity of the major crops," and help small farmers survive in an economic climate that favors the concentration of farmland ownership.

Not only has genetic engineering failed to produce long-promised yield improvements, but the Bank's lending policies continue to concentrate land ownership. Nonetheless, the Bank seeks to minimize the social and economic costs of adopting biotech methods, aid in the development of Intellectual Property Rights (IPR; *i.e.*, patent and licensing) regimes for seeds and plants, offer expertise in risk assessment and biosafety procedures, promote and support research on non-commercial staple food crops, help integrate biotechnology into existing national research programs, and promote public-private partnerships to advance biotechnology.

The implementation of these strategies came into sharper focus following the publication of a 1997 report and a subsequent series of international meetings which crystallized the earlier discussions into the development of a more focused biotechnology strategy.[19] The latter report offered a far more measured and realistic assessment of the state of biotechnology research, balancing this with a broad overview of the known environmental consequences of GE crops. It also proposed a substantially broader agenda, aimed at developing scientific and regulatory expertise to assess and identify potential problems, as well as research on new crop varieties through the existing Bank-supported agricultural research centers in various countries. Numerous consultations and international seminars that followed the report's publication proposed an even broader pro-biotech focus, however, emphasizing the protection of IPR and the development of public-private partnerships promoting biotechnology.[20]

During the same period, the Bank was increasing its collaboration with the agrochemical industry, even while outwardly promoting concepts such as sustainability and Integrated Pest Management (IPM). The Pesticide Action Network (PAN) reviewed Bank documents describing over 100 agricultural projects approved between 1997 and 2000, and found a persistent focus on intensifying production and increasing farmers' access to agrochemicals, despite a 1998 policy emphasizing IPM-based alternatives.[21]

PAN also uncovered an ongoing Staff Exchange Program, through which the Bank had entered into business partnerships with nearly all the leading pesticide companies, including biotech giants Aventis, Novartis, and Dow. "For public monies to support the placement of World Bank staff at these companies," argued Marcia Ishii-Eiteman, coordinator of PAN North America's World Bank Accountability Project, "constitutes a gross violation of the Bank's pest management policy and its business partnership guidelines. It is also antithetical to the Bank's commitment to sustainable development and a misuse of public funds."[22]

The Staff Exchange Program involves 189 corporations, governments, universities, and international agencies, including leading transnational companies involved in agribusiness, pharmaceuticals, petroleum, mining, timber, and banking.[23] The Bank trades staff members with various partner institutions for periods of up to two years, with a provision for adding additional years, and assignments are targeted toward mutual institutional needs. Exchanges with agribusiness/biotech companies were most active during the late 1990s and the beginning of this decade.

For example, an Aventis (now Bayer CropScience) marketing analyst spent nearly four years helping the International Bank for Reconstruction and Development, the largest and most visible Bank division, to develop its position on agricultural biotechnology, as well as strategies for leveraging financing from the private sector through the Bank's International Finance Corporation (IFC). A Dow sales officer worked on projects in Africa and Mexico, and served on teams studying agrochemicals and biotechnology. Novartis' (now Syngenta's) head of public affairs spent a year working on outreach and communications strategies for the Bank's rural development unit. Meanwhile, Bank officials stationed at Novartis and Rhone Poulenc Agro (now part of Bayer) in the late 1990s worked on biotechnology regulatory issues and rural development partnerships.[24] Through these exchanges, the Bank adjusted its biotechnology strategies to better satisfy the leading biotech seed developers, and corporations gained access and influence over public policies in the global South.

BIOSAFETY IN INDIA: A MODEL PROJECT

In 1996, the annual Conference of Parties to the UN's Convention on Biological Diversity (a product of the 1992 Rio "Earth Summit") launched negotiations toward an international protocol on the safe handling and transport of genetically modified organisms (GMOs). African nations, along with others concerned about how future imports of GMOs might threaten the integrity of indigenous plant and animal species, forged an international consensus requiring countries seeking to export intact, viable GMOs — such as live plants, seeds, and microorganisms — to obtain the consent of the importing country. Despite numerous obstacles imposed by the major GMO-producing countries — particularly the US, Canada, and Argentina — a full text was developed in Cartagena, Colombia, in 1999 and approved in Montreal in January 2000. The Pacific Island nation of Palau became the 50th country to ratify the Cartagena Protocol on Biosafety in June 2003, bringing it into effect following a 90-day waiting period. As of March 2004, 89 countries had ratified the protocol.[25]

The Biosafety Protocol requires that countries seeking to export living modified organisms (LMOs) obtain "prior informed consent" from the importing country. Where organisms are intended for introduction into the environment (*e.g.*, seeds), detailed information on their identity, traits, and characteristics needs to be communicated, and the receiving country may invoke the Precautionary Principle[26] in deciding whether to allow the import. In the case of organisms intended for contained use, such as in laboratories, the exporter only needs to label the LMOs and specify rules for their safe handling and use.

GMOs intended for food, feed, or processing are largely exempt, but are required to carry the label, "May contain living modified organisms," as well as a certification that they are not intended for environmental release. Pharmaceutical products regulated by the World Health Organization are exempted entirely. Countries that have not ratified this protocol — including the US, which hasn't signed the Biodiversity Convention and, therefore, is not eligible to do so — are expected to trade with signatory states in a manner that is consistent with the document's objectives, though there are no means to enforce this.

In March 2003, the World Bank approved detailed plans for a three-year, $3 million dollar project designed to help India fulfill its obligations under the Cartagena Biosafety Protocol. The project is designed to enhance the capacity of various government agencies and research centers to implement the agreement's provisions. In partnership with India's Ministry of Environment and Forests, the Bank will help build "technical capacity for risk assessment, management and monitoring"; establish the required database system and clearinghouse mechanisms for GMO imports; support the development of infrastructure for research, risk assessment, and monitoring; and strengthen laws, regulatory frameworks, and "communication strategies."[27] Some 1600 individuals will participate in a series of risk evaluation courses, with primary support for four existing biotech research centers in India.[28]

The Bank's Project Brief reviews in some detail the agencies and facilities that are currently engaged in biotechnology regulation and research in India, and addresses ways to significantly expand the capacity of these institutions. The Bank assumes that, having ratified the Cartagena Protocol, India will inevitably see an "increased movement" of GMOs into and within the country.[29] The proposal postulates an ever-growing need for researchers to identify and monitor laboratory research as well as field trials; ensure safe handling of GMOs; evaluate environmental risks such as pollen transfer and affects on "non-target" species; assess safety procedures and guidelines; evaluate the safety and nutritional composition of engineered foods; and monitor and regulate commerce in engineered organisms.[30]

An ever-expanding array of scientists and public officials will engage in the detection, tracking, and evaluation of GMOs. While some researchers may shift their priorities from the development of new GE organisms to the evaluation of their safety, this project entails a significant expansion in the capacity of Indian researchers to work with GMOs and promote their "societal acceptance."[31]

Dr. Klaus Ammann, a Swiss biotechnologist enlisted by the Bank to evaluate this project emphasized the desire to "avoid unwelcome frictions in the so-important process in modernizing [India's] agriculture," and raised a concern as to whether this project will be implemented "expediently enough."[32] He suggested that the work in India may provide a model for circumventing the political and regulatory problems that in his view have unnecessarily impeded the introduction of GE products in Europe, and proposed ways to streamline the project to further the acceptance of genetic engineering.

This analysis punctures the balloon of scientific objectivity that surrounds the project, and removes any ambiguities as to its intended outcome. Along with many European countries, India has been one of the world's centers of opposition to genetic engineering and now, with the support of the Bank, will build a rather extensive infrastructure for the evaluation, testing, and development of new GE products. While regulation and testing have become essential to guard against the undesirable consequences of GMOs, the mobilization of this greatly expanded infrastructure could make it politically inexpedient to refrain from introducing an ever-increasing variety of GE products and GMO-related research activities into the country.

A Kinder Biotech Agenda?

Another cornerstone of the Bank's involvement in biotechnology is the Consultative Group on International Agricultural Research (CGIAR). Created by the Bank in 1971, it was co-sponsored by the UN's Food and Agriculture Organization (FAO) and the UN Development Program. It brought together four privately funded agricultural research centers — the International Rice Research Institute (CIRRI) in the Philippines, a maize and wheat research center (CIMMYT) in Mexico, and tropical agriculture centers in Colombia and Kenya — under a single umbrella.[33] The Ford and Rockefeller Foundations were the primary supporters of these four original centers, which were seen as flagship institutions of the "Green Revolution" during the 1960s and Seventies.

Today, the CGIAR consists of 16 research centers, with focuses including livestock, fisheries, water management, and agroforestry. It also supports National Agricultural Research Systems in numerous countries, under the auspices of the International Service for National Agricultural Research (ISNAR), and maintains at least 10 percent of the world's gene banking capacity, housing more than half a million samples of both wild and cultivated plant species. In its first 25 years, CGIAR raised over $4 billion in support of agricultural research.[34]

As the Bank as a whole shifted its research goals during the 1990s toward an emphasis on "poverty reduction" and the conservation of biodiversity, so did CGIAR. "Biotechnology," according to CGIAR's Website, "will be mobilized through research alliances to ensure that it can contribute to more sustainable agricultural growth in developing countries, with special care devoted to issues such as ethics, safety and the access of developing countries to biotechnology products."[35] This often translates into the promotion of biotech research for its own sake.

In a 1998 speech, former CGIAR Chairman and World Bank Vice President Ismail Serageldin proposed a more benign biotech research agenda for these centers. Seeking a "scale neutral" approach to biotechnology that would carry "no intrinsic bias against the smallholder farmer," Serageldin suggested that the center's biotech research focus on "environmentally friendly" approaches to disease and pest resistance, as well as tolerance to environmental stresses such as drought and salt, and the development of new livestock vaccines.[36] To address the safety and ethical concerns that had already sparked widespread debate in Europe and elsewhere, Serageldin suggested that CGIAR focus on "low risk enabling biotechnologies" and traits that "do not involve weediness or a virus component."

This approach would represent a significant departure from the research agenda of Monsanto and the other corporate developers of genetic engineering, where the imperative to introduce new commercial products clearly overrides all other considerations. On the other hand, Serageldin's speech suggests a certain naïveté that risks can be avoided by simple departures such as limiting gene transfers to plants of the same, or closely related, species. The history of GE research demonstrates that the very process of genetic engineering introduces an array of uncertainties and disruptions to genetic regulation, regardless of the source of the introduced genes.[37]

A 2003 review by the World Bank's Operations Evaluation Department assessed CGIAR's progress to date in fulfilling this agenda.[38] Reviewer William Lesser of Cornell University reported that CGIAR has invested some $25 million per year in biotechnology research, ranging from gene mapping and transgenic crops to policy studies, with 11 of the 16 CGIAR Centers involved in some way in biotechnology.[39] Previous reviews by the Group's Technical Advisory Council had, on several occasions, criticized the centers for "underinvesting" in biotechnology.

The most active of the 16 centers is the maize and wheat research center in Mexico (CIMMYT), with seven scientists working in an Advanced Biotechnology Center. A major focus is reported to be marker-based transfer of stress and disease resistance traits — essentially introducing DNA markers as diagnostics to better assess the outcome of conventional breeding experiments. Other emphases are functional genomics (using genetic sequence information to develop a better understanding of gene function) and apomixis technology (inducing seed formation without pollination), a project not without environmental consequences.

CIMMYT is also engaged in a partnership with Pioneer Hi-Bred, the world's largest seed company (wholly owned by DuPont), to investigate genes and metabolic pathways associated with drought tolerance in corn, an effort that may be aimed at more invasive forms of genetic manipulation.[40] CIMMYT, along with two other CGIAR centers, has also enlisted an IPR specialist, conducted an IPR audit, and led the development of a detailed IPR policy.[41]

The IRRI has been involved in the development of the so-called "golden" vitamin A rice — which author and *New York Times* commentator Michael Pollan accurately described as "the world's first purely rhetorical technology"[42] — and has engaged in stress resistance and genomics research. Other CGIAR centers have been involved in enhanced breeding methods, genomics, analyses of genetic diversity, and searches for exotic and novel genetic traits. The two tropical agriculture centers both support projects involving gene transfer technologies.[43] The ISNAR assists developing countries to "increase the effectiveness and efficiency of their agricultural research systems,"[44] and supports an Intermediary Biotechnology Service, which assists with policy and infrastructure development to further the adoption of new technologies.

In the policy realm, World Bank analysts have offered detailed analyses of possible institutional arrangements, licensing regimes for intellectual property, joint ventures, and technology transfer protocols in an effort to strengthen ties between the National Agricultural Research Systems and the private sector. A Bank senior economist and the former IRRI deputy director general described CGIAR as "a potentially important 'bridge' between advanced private and public research organizations and public research organizations in developing countries."[45]

The 2003 review of CGIAR's operations reveals considerable internal controversy surrounding the centers' biotechnology involvements. One important development was CGIAR's 1998 pledge to refrain from using Terminator-like technologies to produce sterile seeds. Evaluator William Lesser describes this decision as "absolutist" and an undue politicization of the technology assessment process. "Moreover," he writes, "such a step can increase future pressure by the vocal anti-biotechnology community to exclude technologies for political, not scientific reasons."[46]

Clearly, politics is appropriate when it favors the aggressive promotion of biotechnology, but not when it imposes limitations on its development and diffusion. CGIAR stepped into far deeper political waters in 2002 when it invited the Syngenta Foundation to join its list of sponsors. Syngenta is the world's largest producer of herbicides and, after Monsanto, the largest developer of GE plant varieties. This invitation led the group's advisory committee of nongovernmental organizations (NGOs) to freeze its relationship with the organization, alienating other major donors and threatening a significant curtailment of the group's research activities.[47]

Meanwhile, there are institutional pressures to devote greater CGIAR resources to the ownership and protection of patents and other forms of IPR. Lesser described two opposing tendencies on a 1998 CGIAR assessment panel, with those who saw biotechnology as central to the centers' future mission seeking greater IPR protections and others favoring an emphasis on participatory research based on local knowledge. "The simple answer to such a future impasse," Lesser argues, "is in the narrower selection of panel members," suggesting a political agenda of simply marginalizing those who are critical of biotechnology.[48]

GREENWASHING CORPORATE AGENDAS

Despite these clear institutional biases at the Bank and its numerous affiliated agencies, many analysts still believe that the Bank represents an alternative to the aggressive, commercially driven promotion of biotech products that is characteristic of the leading biotech transnationals. The more overt promotion of GE and other controversial biotechnologies is left to another organization, directly supported by the biotech companies, which maintains close institutional ties to the World Bank. This is the International Service for the Acquisition of Agribiotech Applications (ISAAA).

Long-time Bank advisor Gabrielle Persley, formerly biotechnology manager of the Bank's Agriculture and Rural Development Department, serves as the director of programs on ISAAA's board, and Clive James, a former CIMMYT deputy director general, played key role in ISAAA's establishment during the early 1990s. While some Bank officials have disavowed any direct link to the ISAAA, the history of institutional ties between the two organizations is quite compelling, and the ISAAA's regional operations in Africa and Asia are based at two of the leading CGIAR research centers. The ISAAA is directly supported by Monsanto, Syngenta, Bayer, Pioneer, Cargill, and other corporate biotech leaders, along with the Rockefeller, McKnight, and Hitachi Foundations, among others. Its stated goal is "to bring together institutions from national programs in the South, and from the private sector in the North, into partnerships to transfer biotechnology applications."[49]

The ISAAA boasts programs in Africa (Kenya, South Africa, Tanzania, and Uganda) and Asia (Indonesia, Malaysia, the Philippines, Thailand, and Vietnam), has initiated projects in Latin America (Argentina, Brazil, Costa Rica, and Mexico), and offers fellowships to scientists engaged in technology transfer activities. Researchers at Genetic Resources Action International (GRAIN) describe the ISAAA fellows as an emerging "advocacy elite," often maintaining a strong and lasting identification with their corporate benefactors.[50] Recent projects include manipulation of virus resistance genes in potatoes and papayas, development of diagnostics for maize diseases, genetic engineering of cassava and sweet potatoes, cell culture techniques for the propagation of commercial tree species, and the assessment of Bt corn technologies for Asia.[51]

The ISAAA's history in Asia, began with a 1996 meeting of its board at the International Rice Research Institute headquarters in the Philippines.[52] Ongoing research and technology transfer projects in Asia, and around the world, follow the general pattern of Bank-funded efforts but are much more heavily tilted toward the development of new transgenic crop varieties. These include efforts to develop GE virus resistant potatoes in Mexico, tomatoes in Indonesia, and papayas in Malaysia.

As with similar Bank-supported efforts, these research priorities reflect little regard for local crop diversity or actual on-the-ground agronomic problems. Large sums are committed to the development of GE varieties that aim to do what local farmers can often accomplish by far less invasive means, while introducing new problems, often more serious than those they are purported to solve.[53] The ISAAA also seeks to help developing countries circumvent the biotech industry's maze of IPR rules by facilitating licensing arrangements that give researchers easier access to new, proprietary technologies. While Monsanto spends $10 million a year suing and harassing US farmers to strictly obey their "technology agreements" and other IPR rules,[54] their goal in the global South is clearly to promote the rapid acceptance of new GE crops at all costs.

The main vehicle for direct links between the World Bank and the private sector is the Bank's IFC, with an explicit role of leveraging financing from the private sector for international development projects. In the late 1990s, the IFC developed a $30 million Biodiversity Enterprise Fund for Latin America, which aimed to encourage "sustainable uses" of that region's biodiversity. The Fund encouraged investment in questionable activities including bioprospecting, elite forms of ecotourism, and establishing tree plantations as "carbon sinks" to relieve the climate burden of methane production from cattle ranches.[55]

Today, more than 15 Biodiversity Enterprise Funds exist worldwide under the umbrella of the Conservation Finance Alliance (CFA). Two World Bank officials serve on the steering committee of this organization, along with staff from the Nature Conservancy, Worldwide Fund for Nature (WWF), Conservation International, and other organizations dedicated to "free market" approaches to environmental problems.[56]

These Funds pledge to provide needed capital to mostly small, locally based enterprises in regions of high biodiversity, while offering investors opportunities to simultaneously satisfy financial, social, and environmental goals. The CFA's examples of "proven sustainable finance mechanisms" include user fees for tourism, payments for bioprospecting, debt-for-nature swaps (trading small portions of a country's debt for reallocation of particular lands for conservation),[57] conservation trust funds, and "carbon investment projects" (forestry projects aimed at mitigating industrial carbon dioxide emissions.)[58] Many of these measures have indeed reaped financial rewards for investors — and public relations points for cooperating NGOs — at the expense of indigenous peoples and the ecosystems upon which they have traditionally relied.[59] The CFA's online guidebook cites some seemingly admirable projects: encouraging small farmers in Central America to grow organic cocoa, for example. But it also promises returns on investments of up to 30 percent,[60] a goal that appears quite incompatible with the mission of supporting human-scaled, ecologically sustainable practices in rainforests and other fragile ecosystems.

In an investment climate where corporations routinely shut down plants that fail to produce ever-increasing profit margins, and "socially responsible" investment funds sustain their bottom lines by investing heavily in Coca-Cola and Microsoft, the goal of encouraging conservation through investment appears to be yet another capitalist fantasy. As corporations and the largest international NGOs collaborate in ever-more elaborate schemes to sustain the myth that extracting profits and sustaining the environment can be made compatible, it is clear that the World Bank will continue to play a key role in advancing and legitimizing this dubious agenda, at the behest of global agribusiness and other corporate interests.

Biotechnology and Food Security:
Hard Choices for Africa

LAWRENCE TSIMESE

ROM ZAMBIA TO GHANA, AFRICAN COUNTRIES HAVE WIDELY differing attitudes toward the application of biotechnology to food production. In a context of systemic inequality, the process raises key issues of good governance and global justice as well as science. Can this new technology be used to address poverty and advance sustainability, or will it merely increase global corporate control?

The furor and drama that unfolded in the aftermath of the genetically-engineered (GE) food donation to southern Africa in 2002, and its subsequent rejection by Zambia, evoked memories of decades past. From colonial times, Africa has been the dumping ground for all sorts of discredited ideologies, technologies, and economic policies.

My country, Ghana, is no exception, but in the case of biotechnology, we seem to be taking a different approach. This was exemplified at a press conference held in 2003 by the Science and Technology Policy Research Institute (STEPRI) to shed light on the goals, targets, and activities of the Institute's Biotechnology Development Program. STEPRI's senior scientific officer, George Essegbey, called on the country to explore and harness biotechnology in order to obtain desired benefits, especially in the areas of agriculture and health. If developed and managed efficiently, Essegbey claimed, biotechnology could improve agricultural production, add value to natural products, diagnose and prevent diseases, conserve biodiversity, and preserve the environment.

Advances in biotechnology have indeed opened up a wide range of application opportunities in the developing countries, especially in the health and agriculture sectors. However, some of these advances, such as genetically modified (GM) foods and animal cloning, are controversial everywhere, and some southern African countries have boycotted GM grains in the midst of hunger because of safety concerns.

BIOSAFETY: THE AFRICAN CHALLENGE

It is often argued that agricultural biotechnology holds great promise for Africa. Tissue culture and marker-assisted selection — the use of DNA probes to facilitate traditional plant breeding — are already in widespread use across the continent while, for most countries, genetic transformation is still in the developing stages. The safe application of these technologies requires functioning biosafety systems, and it is a matter of consensus that improving food security and agriculture requires more than technology. Good governance, wise policies, infrastructure, and investment are other key requirements. While African countries have made impressive progress in biotechnology and biosafety, such undertakings obviously require funding, national commitment, and political will. Africa faces particularly high hurdles in all these areas.

Agricultural and social systems in Africa differ considerably from those in the West. There are clear differences in both the approach to and emphasis placed on biosafety. First, while many smallholder farmers have adopted the use of (non-GM) hybrid seeds, others still rely on saving seed from the previous harvest to plant in the next growing season. The right of farmers to save seed is probably one of the biggest issues in risk management, since the practice makes it almost impossible to specify and monitor the conditions of use. In some sectors, genetically modified organisms (GMOs) are identified with the Terminator technology — which has not yet been commercialized — leading to fears that farmers will become dependent on large seed companies. While biotechnologists claim that Terminator-like technologies will help solve the problem of pollen drift, it remains highly uncertain and clearly cannot be recommended under these circumstances.

Second, many of the most important crops in Africa, such as bananas and the root and tuber crops (cassava, sweet potato, potato), are not normally grown from seed, nor supplied through seed companies. Some farmers in countries such as Zimbabwe and Kenya currently obtain seedlings grown from virus-free tissue cultures, but informal propagation will always occur. Such a scenario creates a challenge for the biosafety framework. Once a genetically modified cultivar of sweet potato is released into the market, it will undoubtedly spread to other areas.

The biotech industry claims that Africa needs biotechnology to lower food costs, help feed the poor and hungry, save the environment, and lower pesticide and herbicide use. However, the true motives for genetically altering our food crops appear to be quite the contrary. There is already a mountain of evidence that there are severe flaws, both in the technology and the claims made by the biotech industry. For example, bacterial, viral and fungal genes are now routinely spliced into several of the world's most widely used agricultural crops, with largely unknown consequences.

THE RISKS OF TECHNOLOGY

Some scientists fear that biotechnology may prove more destructive than nuclear weapons. Dr Mae-wan Ho of the Institute for Science in Society says: "The large-scale release of transgenic organisms is much worse than nuclear weapons or radioactive wastes, as genes can replicate indefinitely, spread and recombine."[1]

Dr. John Fagan of the Institute of Science, Technology and Public Policy echoes that conclusion: "We are living today in a very delicate time, one that is reminiscent of the birth of the nuclear era, when mankind stood at the threshold of a new technology," he notes. "No one knew that nuclear power would fill our planet with highly toxic radioactive waste. We were so excited by the power of a new discovery that we leapt ahead blindly, and without caution. Today, the situation with genetic engineering is perhaps even more grave because this technology acts on the very blueprint of life itself."[2]

When GMOs were first released into the environment the biotech industry said we had nothing to fear from these new genetically manipulated foods. They claimed the nutritional content was not changed and that no special labeling should be required. They also said the altered DNA is easily digested and broken down in our gut and does not interfere with our bodies' biochemical makeup in any way. However, no long-term studies have been conducted on human beings to determine conclusively that biotech foods are safe. Nobody knows for sure whether our immune systems are being harmed. We are consuming DNA and proteins from viruses, bacteria, and fungi that have never been eaten by humans in all the millennia we have existed on this planet.[3]

A panel of biotech experts from six African countries — Uganda, Ethiopia, Kenya, Tanzania, Malawi, and Ghana — has argued that it is misleading to confine discussions of biotechnology to genetic engineering, especially GMOs. Biotechnology in Africa overwhelmingly involves tissue culture and related activities. However, the experts,

69

who met in Nairobi in May 2002, sought to ensure that the continent acquires the skills and facilities to deal with all aspects of gene manipulation. It seems that most African countries want to seek a middle ground in the biotechnology debate, avoiding any clear policy regarding the introduction of this technology into their individual countries.

At the same time, resistance to genetic engineering is rising. For example, in January of 2000, Ghana's ex-Minister for Environment, Science and Technology, Cletus Avoka, assured the nation that his ministry was doing everything possible to check the influx of GM foods. Answering a question from Parliament, Avoka noted that the GM food issue has generated controversy in developed countries and declared that Ghana, a developing country, would not risk allowing its use. On the Terminator seed technology, he said the government would not tolerate its use since it has the potential to harm crop management. He said his ministry was participating in the development of a global Biosafety Protocol (see Chapter 3), which, among other things, would insulate the world's biodiversity from the possible adverse effects of GMOs.

There are further indications of resistance to genetic engineering, as governments worldwide are adopting mandatory labeling, rigorous testing, and other restrictive measures over GE products, including banning GE products where the evidence warrants. In Africa specifically, a group of nations led by Ethiopia is developing legislation that would make it illegal to import GE foods or crops to their countries without prior approval. This prior consent law is designed to press GE exporters to carry out human safety, environmental, and socioeconomic studies. Of course, this initiative has drawn opposition from biotechnology corporations and grain-exporting nations, led by the US, who consider all biosafety measures a restraint of trade.

Meanwhile, a concerted and well-coordinated effort is being made across Africa to gain entry for GMOs. In the face of tight regulations, organizations that promote biotechnology are looking for ways to co-opt the concept of biosafety for their own benefit and are calling for "harmonizing" or equalizing biosafety regulations across Africa and the South. Reading between the lines, it is clear that they are seeking to water down the rules. However, countries in the former Organization for African Unity, now the African Union, are moving in a different direction. Concerned that the Cartagena Protocol on

Biosafety does not go far enough in protecting their biodiversity and agriculture interests, they have written a much stronger Africa Model Law on Biosafety.

BIOTECHNOLOGY AND SUSTAINABILITY

In contrast to the European Union, most African countries are not absolutely rejecting GMOs but are attempting to tread a path between this new powerful system and the status quo. This is understandable because of the chronic struggles with hunger in Africa, and the pervasive idea that hunger may be an epidemic. Most nations are careful not to throw away any opportunity to solve the hunger problem once and for all, but at the same time they would like to protect their environment and food safety. Thus, they are proceeding carefully and ethically.

Many scientists and policy makers are making statements such as this: "For the benefit of humankind, we must end the squabbling over biotechnology and allow objectivity to prevail. Its potential to feed the hungry, heal the sick and make life better for billions of people is too great to lose to fear and confusion."

But despite this tantalizing picture of what biotechnology might do for the starving millions in Africa and other parts of the Third World, there is growing awareness that genetic engineering is a central means by which global capitalism is consolidating its control over our food and health care. Biotechnology has helped drive unprecedented corporate concentration in both agribusiness and the pharmaceutical industry. Many of these corporate entities, particularly the agribusiness giants, have concluded that Africa lacks the technological expertise to meet its own food requirements, and are doing everything possible to lure governments into adopting GE technology as a panacea to end hunger and stimulate economic prosperity.

GE crops, however, promote monoculture. In Africa and other developing countries, farmers successfully control pests by encouraging biodiversity in their fields and encouraging beneficial insects and crops. The UN Food and Agriculture Organization points out that more plant diversity has been lost because of industrial agriculture than any other cause. GE crops will only increase this problem.

Scientists have shown that reductions in biodiversity lead to the evolution of aggressive pests and diseases that are more difficult to control than those from which they were derived.[4] Millions of farmers in developing countries rely on farm-saved seeds for their crops, but once they begin to buy GE seeds they will be dependent on future purchases, as Monsanto and other companies prohibit seed saving.

The fact that such a technology is largely in the hands of the private sector in the North can lead to biases in the type of research that is conducted. It is only logical that large companies would aim at worldwide markets for their products, but these products might not be appropriate for small farmers in developing countries, who tend to work in highly variable and vulnerable ecosystems and need seeds that are location-specific.

In the end, biotechnology could undermine food security in Africa rather than secure it. If the biotech corporations want to feed the hungry, they should encourage sustainable land reform that puts farmers back onto the land, and also push for wealth redistribution that allows the poor to buy the food of their choice.

"Beggars Can't Be Choosers:"
GE Food Aid and the Threat to Food Sovereignty

SHIRI PASTERNAK

"To those who have hunger, give bread. To those who have bread, give a hunger for justice."

— Latin American Prayer

WHEN DR. TEWOLDE EGZIABHER, HEAD OF ETHIOPIA'S Environmental Protection Agency, explained genetic patenting and the biotech seed industry to farmers in Ethiopia, they had a hard time understanding how anyone could keep seeds from those who needed them. "Their tradition is if you want some seed from what they have, take it."[1]

In modern capitalist societies, however, the patenting of "innovation" is an engine of the political economy. The fiscal value of knowledge is even woven into our language. To verify something, for example, it is not uncommon to turn to someone and ask: "Do you buy that?" Intellectual property rights are embedded within a culture of utility and objectification. More than just a means of capital accumulation, the ownership of ideas — and now life itself — is a tool to comprehend reality, and also to produce, control, and normalize it. One way that this power of ownership and control is reproduced is through charity.

North Americans generally consider food aid to be a politically neutral, charitable project. Images come to mind of flood victims floating on rooftops or fleeing masses from a war-torn countryside being offered food, shelter, and clothing. Although charity is ostensibly intended to relieve social and economic inequality, it also reflects broader relations of power. Charitable acts have the power to redefine the problems being addressed, while a critical view of charity, as well as science, technology, and their social consequences can expose the interplay of interests and the underlying structures of power.

The biotechnology industry claims to be developing technology that will "feed the world." However, much like the Green Revolution, the new model of the "gene revolution" does not address the complex, underlying issues behind food scarcity. It is at least equally dependent on costly fuel, chemicals and, now, patented seed. The biotech giants seek *dependents*, while the US government needs an easy way to dispose of the surpluses created by its heavily subsidized crops. Foreign aid programs create markets that satisfy these overlapping interests.

Dumping genetically modified (GM) food into vulnerable markets carries with it the power to disrupt agricultural economies and knowledge systems around the world, increasing food insecurity and making the goal of a GM-free agriculture appear increasingly remote. The US biotech industry, largely through the US Agency for International Development (USAID) and the UN World Food Program (WFP), is dressing up worldwide GM contamination in a philanthropic disguise. This aid model offers far more than mere sustenance. It's all about Western control. As one US official put it: "Beggars can't be choosers."[2]

A BRIEF HISTORY OF US FOOD AID

The history of American food aid is a story of political carrots and sticks. Since World War II, US food aid has played a significant, often invisible role in shaping the new agrarian order. In 1950, developing countries accounted for 10 percent of wheat imports; by 1980, they were receiving 57 percent.[3] This dramatic increase is directly attributable to the fact that US aid promotes a Western diet and undercuts local food programs. Globally, from the postwar years until the 1970s, the US provided 90 percent of all food aid, although this figure has since dropped to between 60 and 70 percent.[4] It is also safe to assume, according to USAID, that at least 30 percent of the 500,000 tons of maize it exports is genetically modified.[5]

Since aid agendas reflect the political and economic interests of changing US administrations, food aid plays many roles: a bargaining chip in peace negotiations (the Middle East accords); a way to gain military favors (Pakistan); long-term support for an ally (Israel); continuing support for staunchly capitalist regimes (El Salvador, Costa Rica, Honduras), and a means to ensure geopolitical strongholds for the "war against terror" (Uzbekistan).[6] However, as much as these regimes may differ, US corporations have reaped the benefits. Direct state oppression is not always a necessary form of social control, particularly when it can be managed equally well through economic measures.

Following World War II, US food aid played a Janus-faced role as both humanitarian assistance and a means to dispose of crop surpluses. By 1954, the US had donated over 18 million tons of wheat — much of it grown under price supports created after the Depression — to Europe through the Marshall Plan. With Europe's production strength rebuilding, however, the US needed new markets for agricultural and other products. In order to bolster its position in world trade, the US devised a new program known as "Food for Peace." It sought to expand foreign markets for US agricultural products, combat hunger, and encourage economic growth in developing countries.

Under Food for Peace, food aid could be offered under four major programs, but approximately 70 percent were so-called Title 1 sales. These "sales" provided US agricultural commodities to "friendly" (capitalist) nations in the form of credit. Usually, developing countries repaid these loans by depositing their own currency into local bank accounts. By the late 1960s, the US government had amassed several billion dollars in foreign currencies by this means. Between 10 and 15 percent was then loaned to US corporations for the explicit purpose of generating profitable economic activities inside recipient countries. As a result, 419 subsidiaries of US firms were established in 31 countries.[7] This era marked the rise of the transnational agribusiness corporation.

In the 1970s, facing a debt crisis, insatiable market demands for agricultural commodities, and sudden competition with Europe and Japan for markets, US-based creditors came calling. The US demanded immediate repayment of Food for Peace loans and other bilateral debts, amounting to billions of US dollars. When developing countries couldn't pay their debts, exacerbated by dwindling or fluctuating markets for prime resources, the World Bank and International Monetary Fund (IMF) stepped in to offer new loans. Conditions for lending, through the now-notorious structural adjustment programs (SAPs), often included selling off surplus grain and opening up markets to cheap food commodities that flooded in at prices lower than the cost of growing them locally. Nations that had been self-sufficient for centuries became increasingly reliant on food aid. As Evelyne Hong reports on Somalia:

> "Until IMF-WB intervention in the early 1980s, agriculture
> in this country was based on reciprocal exchange between
> nomadic herdsmen and traditional agriculturalists. In the

1970s commercial livestock was developed and this affected the nomadic herdsmen. Until 1983, livestock contributed to 80 percent of export earnings. Despite recurrent droughts, Somalia was virtually self sufficient in food until the 1970s. From the mid 1970s to mid 1980s, food aid increased fifteen fold at 31 percent per annum. The influx of cheap surplus wheat and rice in the domestic market soon displaced local producers and caused a shift in food consumption patterns to the detriment of traditional maize and sorghum."[8]

Food aid not only flooded foreign markets with Western commodities, it displaced the local agricultural economies of hundreds of communities. This practice of "food dumping" is among the most violent impacts of the "free trade" ideology. For example, in the Philippines in 1998, 15,000 metric tons of corn rotted on the island of Mindanao due to low market prices, while 462,000 tons of corn flowed in from the US under the Food for Peace program.[9] Efforts have been made to ameliorate the impacts from food aid, but the regulatory system put in place by major international institutions is fragmented and incoherent.

In recent years, the WFP has become the biggest cog in the US international food aid machine. Established in 1963 to provide food aid through a variety of UN programs, it was the first multilateral program introduced to deal exclusively with food aid.[10] The WFP is a UN agency, run in cooperation and coordination with various non-governmental organizations, and administering emergency and development projects in 83 developing countries. It is the single largest source of "grant aid" within the UN system, but it relies on voluntary contributions to finance projects. The principal sources of funding are governments — particularly the US, Japan, and the European Union — but businesses and individuals are also contributors.[11]

In 2002, the WFP admitted that it had been distributing food "with some biotech content" around the world for at least seven years.[12] Although operating in 83 countries, it made no effort to notify recipients, including India, Columbia, and Guatemala, of this fact. The WFP defends its actions by pointing out that it relies on the UN's food safety guidelines — the Codex Alimentarius codes — which are also the US and WTO standards. Codex has entertained several specific health-related challenges to GE, but generally bases its rulings on the narrowest interpretations of available scientific evidence. (see Chapter 2).

An international agreement known as the Food Aid Convention (FAC) recognizes that food aid should be culturally acceptable, bought from the most cost-effective source available, and, if possible, purchased locally so that regional markets do not suffer. The guidelines, first established in 1967, have been modified several times since, most recently in 1999. The US is a signatory to the Convention but perpetually violates its guidelines,[13] underscoring a need for more than just procedural protocols.

BIOTECH AND FOOD AID

Since 1996, increasing quantities of US agricultural products have been genetically engineered (GE). Each year, more than two million tons of food from genetically modified organisms (GMOs) are sent directly to developing countries by US foreign assistance agencies such as USAID.[14] In addition, the WFP distributes another one and a half million tons of transgenic crops donated by the US.[15] The US now provides more than 60 percent of WFP's total aid intake.[16] Unlike most other Western governments, which give money, the US insists on donating in-kind commodities or tying cash contributions to purchases of US produce. Transgenic aid is sent unlabelled *in spite* of countries' biosafety protocols and legislation, casting a shadow of doubt on labeling campaigns. As Naomi Klein writes: "The real strategy is to introduce so much genetic pollution that meeting the consumer demand for GM-free food is seen as not possible. The idea, quite simply, is to pollute faster than countries can legislate — then change the laws to fit the contamination."[17]

It is not difficult to trace the relationship between the US government and corporate biotech proponents and their ability to influence the WFP, which relies on donations and is therefore often captive to donor agendas. Since 1961, USAID has been "the principal US agency to extend assistance to countries recovering from disaster, trying to escape poverty, and engaging in democratic reforms."[18] Much of this food is funneled through the WFP. But USAID does not work like a conventional aid agency. It is more like a public relations front and marketing board for agribusiness corporations. Some examples will help illustrate these relationships.

First, there are USAID's programs of persuasion. One is the

Collaborative Agriculture Biotechnology Initiative (CABIO), funded to lobby "for stricter intellectual property rights legislation and plant variety protection in developing countries."[19] CABIO and other organizations pressure foreign governments to adopt biotechnology and its attendant package of laws and regulations. To the same effect, over the past 10 years, USAID has paid biotech companies to run projects in the developing world with the cooperation of local research institutes. These include collaborations between Monsanto and the Kenyan Agricultural Research Institute to develop virus resistant sweet potatoes; Garst Seeds and Indonesia's Central Institute for Food Crops to develop tropical Bt maize; and Pioneer Hi-Bred and the Egyptian Agricultural Genetic Engineering Research Institute to further develop Bt pesticide technology.[20] Research in developing countries is a stepping-stone for corporations to begin test-growing crops and influencing agricultural production. With the help of USAID-funded lobby groups like CABIO, the legal biosafety protocols can be written by the time the research is done. In this way, the research is protected and new legal markets are created for biotech products; the science is easily controlled and normalized.

Another form of "aid" comes from institutions that influence social and political elites to support GE, and integrate the technology into the country's intellectual and scientific discourse. The International Service for the Acquisition of Agribiotech Applications (ISAAA: see Chapter 3) is just such an organization. It works by building "global partnerships" between Northern corporations and the public sector in the South.[21] The ISAAA has set its sights on 12 countries in Asia, Africa and Latin America: Indonesia, Malaysia, the Philippines, Thailand, Vietnam, Kenya, Egypt, Zimbabwe, Argentina, Brazil, Costa Rica, and Mexico. Devlin Kuyek, a researcher with Genetic Resources Action International (GRAIN), sums up the organization's ultimate aims: "All ISAAA projects serve primarily to awaken interest in and commitment to biotechnology within national agricultural research systems and to develop national capacity to conduct biotechnology research and development."[22] USAID is one of its main funders.

One final USAID-funded biotech organization deserving mention is the African Agricultural Technology Foundation (AATF), based in Nairobi, Kenya. It was set up "to remove many of the barriers that have prevented smallholder farmers in Africa from gaining access to existing agricultural technologies that could help relieve food insecurity," and conceived by the Rockefeller Foundation, a backer of ISAAA and instrumental financier of the Green Revolution.[23] This effort is also funded by four of the world's largest seed/agrochemical companies. "The foundation's aims are to identify crop problems in Africa that might be amenable to technological solutions. It then plans to negotiate with the corporations involved for assistance and patent licenses and seek support from African governments to help put new subsistence farmers across the continent."[24] AATF staff have assured the public that they'll donate this technology when the research is finished, but fear of corporate control of seeds and contamination, coupled with the expectation of an inevitable sales pitch down the road, leave many unconvinced.

OVERPRODUCTION, SURPLUS, AND DUMPING

GM crops may have cost the US economy $12 billion in 1999-2001 alone, due to international protests against the technology.[25] One report states, "Since 1997, the European Union (EU) has virtually ceased importing corn from the US, exports dropped from more than 1.5 million metric tons in 1997 to less than 70,000 metric tons in 2000. US exports of soybeans to Europe fell from 11 million tons in 1998 to 6 million tons in 2000. An Iowa State University economist calculated this to be the equivalent of losing a market for one out of every three bushels of soybeans grown in Iowa."[26]

Nevertheless, production of GE crops in the US is expanding. ISAAA reports that these crops increased by 12 percent, or 6 million hectares, in 2002. GRAIN researchers respond by pointing out that "the area grown to GM crops around the world in 2002 was 58.7 million hectares, which is still less than two percent of the total agricultural land around the world." [27] More than 99 percent of the GE crops planted globally in 2002 were in only four countries — the US, Canada, Argentina, and China — and the US is responsible for nearly two-thirds of all GE plantings. With GE acreage growing faster within the US, food aid has become a primary means for addressing the paradox of a decline in markets accompanying growth in plantings.

The 2002 US farm bill provided a record $180 billion over 10 years in subsidy payments to farmers, almost exclusively for commodity crops such as maize, corn, soy, cotton, wheat, and rice. [28] But even before the bill was introduced, farmers had become dependent on emergency support, including assistance to compensate for market losses, which increased dramatically just as GE crops were coming into wider use in 1998.[29] Expanded trade was also a contributing factor, but farm analysts point out a high correlation between the acceptance of GE crop varieties and increased levels of assistance for farmers experiencing loss of markets.

During the 2001 budget cycle, US President Bill Clinton approved a foreign operations bill that included substantial Development Assistance funds for the expansion and export of biotech products. The bill allocated $310 million for USAID work in Central and Eastern Europe and in developing countries for agriculture and rural development; $30 million was specifically designated for biotechnology research and development aimed at addressing the developing world's environmental, humanitarian, and health concerns.[30] A Greenpeace spokesperson in India commented that "if they can't lever this technology into the South, the whole project is dead."[31]

Groups in Eastern Europe are outraged at being the targets of this promotion: "In the Ukraine, we already have to live with the legacy of Chernobyl. In the last decade, we have become the dumping ground for nuclear technologies. Now, we see the transfer of another hazardous technology, unwanted in the west — ag-biotech," said Tamara Malkova, from the group Green Dossier, based in Kiev, Ukraine.[32]

Instances of unwelcome GE aid have been recorded in Central and Eastern Europe, Latin America, Africa, and Asia including Georgia, Sri Lanka, the Philippines, India, Bosnia, and Burundi. One recent case involved Albania. Sixteen thousand tons of maize arrived there in October 2003 as part of a US-sponsored "Food for Albania" program. Demonstrations were held to protest suspected GE shipments, just as non-governmental organizations (NGOs) were trying to secure a five-year moratorium on GMOs. In the same region, Serbia and Montenegro have implemented a GE-free agricultural policy "with a comprehensive law regulating the conditions for deliberate release and placing on the market. However, its GE-free status is threatened by smuggling of GE soybeans, field trials and US food aid donations to Kosovo," maintains Food First.[33] Contamination is a serious concern in the region because Romania and Bulgaria grow commercial GE crops, and there is alleged seed smuggling across borders.

GM FOOD AID TO AFRICA

In 2002, Zimbabwe, Malawi, Zambia, Mozambique, and other

countries in Southern Africa rejected GE food aid from the US. Headlines like "Zambians starve as food aid lies rejected"[34] alerted the public to the alleged inhumanity of African leaders. How could they let millions die by rejecting GE food, just to satisfy European importers seeking GE-free agricultural goods from Africa? When President George W. Bush addressed the African drought at the annual biotechnology industry convention in 2003, he argued: "For the sake of a continent threatened by famine, I urge the European governments to end their opposition to biotechnology."[35] The US counseled Europe to recognize the humanitarian disaster that it claimed would ultimately leave blood on European hands. This provided an ideal opportunity for Bush to take the moral high ground, just as the US prepared to argue the case against Europe's GE ban at the World Trade Organization (WTO).

US officials crafted a remarkable story, one designed to silence Africans and to cast White Europeans as the lone voices of opposition. Lost in the ensuing war of words were the facts that African protocols against GE pre-dated Europe's *de facto* GE moratorium, and that the draft protocols of the Africa Group are actually tighter than the Cartagena Biosafety Protocol (described in Chapter 3). According to Ethiopia's Dr. Tewolde Egziabher, "We took our position long before Europe took this movement to keep GE out, and therefore it is totally unfair to say that the African position is influenced by Europe."[36]

Africans have substantial reasons of their own to reject the aid. Take the case of Bt maize. Officials swear that it is eaten daily by US citizens with no negative effect. But Dr. Egziabher compares the US daily intake — two percent or less of processed corn, mostly in breakfast cereal — with countries where maize makes up nearly 100 percent of the diet. In this sense, Africans will become guinea pigs in an absurdly wasteful and potentially hazardous corporate experiment. Some countries, such as Zimbabwe, agreed to accept the aid if the grain were milled, but the cost is fantastically high compared to the price of non-GM grains that could be obtained within the region.

Beside the human health risks, the environmental uncertainties of accidental crossbreeding between GM crops and regional seeds highlight this clash of worldviews. Egziabher explains: "Most farmers are not really aware of GM until we explain it to them. They get horrified because they cannot imagine that what God made would be mutilated in this way."[37]

It is also critical to refute the argument that Africa's food problems are caused by an inability to produce food. "It is bigger than that," Egziabher insists. "It is in our ability to store and transport food, and process and make it available where it is needed, when it is needed. It is not the technological fix at the variety level that will change it."[38] He points to the economic violence of neoliberal globalization: for example, Malawi had enough stored food to survive a recent drought, but the World Bank insisted that it had to be sold to pay down the country's debt.[39] He and other African authorities emphasize the urgent need for investment in infrastructure and diversification of the rural economy.

Catastrophes like the Malawi drought are played out on a stage set by terrorism, to shame people into believing the pro-biotech hype: "The world is standing by as southern Africa may experience more deaths every day than all those lost on September 11," said Professor James Ochanda, chairman of the African Biotechnology Stakeholders Forum based in Nairobi.[40] Imagine if 9/11 happened to Americans every day, Ochanda urged, invoking the stereotypical CNN-version of September 11: an event devoid of historical causality or broad culpability.

Elsewhere in Nairobi, just three months earlier the African Diversity Network, a grassroots NGO, presented a radically different picture of security offered in the findings of a Seed Security Study conducted in Ethiopia, Kenya, Malawi, South Africa, Uganda, Zambia, and Mozambique: "Any strategy to attain food security ... needs to address the key issues around seed — access to and availability of seed, sustainability of the means of production, cultural and ecological diversity inherent in agriculture and the independence of farmers."[41]

There are few indications that the global rejection of GE food aid has taken the force out of US steamrolling. Biotech company stocks are dropping, but the US administration doesn't hesitate stooping to new lows — namely, using the sick as GE bait. As US Health Secretary Tommy Thompson told Reuters, grants from the $15 billion pledged by the US to fight HIV/AIDS in developing countries contain a few contingencies. Among these, Thompson said, was that Zambia must "re-think its decision" to refuse GE food aid.[42] In March of 2004, USAID provoked a diplomatic row by cutting off aid shipments to Sudan, after the country asked that imported US commodities be certified as GE-free.[43]

NGOs: Subcontracting Humanitarian Aid

In August 2002, Andrew Natsios, director of USAID, accused environmental groups of "endangering the lives of millions of people in southern Africa by encouraging local governments to reject GE food aid. They can play these games with Europeans, who have full stomachs," Natsios said, "but it is revolting and despicable to see them do so when the lives of Africans are at stake."[44]

Africa became the leading recipient of food aid and humanitarian assistance in the 1980s. Simultaneously, NGOs became increasingly important in the management and targeting of this aid. "In net terms, NGOs now collectively transfer more resources to the South than the World Bank," asserts international relations analyst Mark Duffield.[45] Where neoliberal institutions and so-called independent NGOs have coalesced, we can see the face of "what amounts to the West's regional policy for Africa."[46]

Duffield writes that a two-tiered neoliberal system of social welfare has emerged in Africa. First, there are the World Bank and IMF Structural Adjustment Programs, which encourage export-oriented models of economic growth and security. Then there are the NGOs that have emerged to deal with the fallout from these capitalist models. He laments, "The growth of official funding channeled through NGOs, reinforced by the high costs of relief workers, has given donors a significant measure of influence over welfare priorities as the safety-net system has spread."[47] Thus, NGOs are transformed from independent bodies to tools of donor policy, and the politicizing of humanitarian aid is hidden behind their supposedly neutral positions.

Duffield distinguishes between progressive and conservative NGOs, pointing out that the progressive ones are critical of donors, try to remain independent, and expose the inadequacies of aid programs. "However the sheer scale of impoverishment means that NGOs are overstretched, under-resourced and apart from political obstacles, frequently face major logistical constraints."[48] Neither form of social welfare — the World Bank/IMF's or the conservative NGOs' — can respond to the underlying causes of humanitarian crisis, and both are radically disruptive forms of Western intervention.

Recent US relief aid to India's eastern coastal state of Orissa, devastated by a cyclone, consisted of a soy-corn blend that showed GE contamination in tests by the firm Genetic ID. The food was routed through CARE and the Catholic Relief Society, which receive food directly from the US. Despite India's ban on importing GE foods, they have also made their way into the country through the Integrated Child Development Scheme, which used CARE India to feed preschool children and expectant mothers, as well as through the WFP.[49] Greenpeace, along with Vandana Shiva's Research Foundation for Science, Technology and Ecology, attempted to block US food aid to India during this natural disaster.[50]

POWER IN A CONTAMINATED SEED

A number of citizens' juries have convened around the world to deliberate the subject of GE seed and its impacts on food sovereignty, among other related issues. In the Indian state of Andhra Pradesh, citizens put scientists and politicians on trial before a jury of rural women, and heard the testimony of hundreds of peasant farmers. In Zambia, the GE debate extended over four days (see Chapter 6), and 43 participants from 16 districts gathered to consider issues affecting Zimbabwe's small farmers. As Elijah Rusike of the Intermediate Technology Development Group and based in Zimbabwe writes, these assemblies are important tools for "aligning policy decisions with real world situations — in this case, the lives and concerns of small farmers."[51]

In the US, donor prerogatives aimed at global control of agricultural markets carry far greater weight than the needs of the hungry. Realizing the world depended on the US for food aid, the CIA claimed, in a 1974 report for the World Food Conference: "This could give the United States ... an economic and political dominance greater than that of the immediate post-World War II years." An unidentified analyst added: "Washington would acquire life and death power over the fate of the multitudes of the many."[52]

This overwhelmingly militarized power has taken on a new meaning with the introduction of biotechnology. The patenting and ownership of biotech seeds confers a chilling literacy to the classical Marxist critique of capitalism: Today, the "reproduction" of social relations is emerging from deep within the biological structures of altered living cells. By controlling the production and germination of seeds, agribusiness corporations are seeking to monopolize the largest and most pervasive economy in the world: the raising and cultivation of food.

Our knowledge of the intermingling, constitutive histories of natural and human cultures — a history often embedded in the seed — is being flattened into the singular image of a high-tech laboratory. It is not simply that the seed has been turned into a commodity, nor just that it has been genetically altered. The cultural practices and knowledge systems that form — and are formed by — agricultural economies produce unique, locally appropriate crop varieties through meticulous seed selection, human and ecological migration, and adaptation to weather and disease. As models of mutual interdependency, these practices also offer an ethical framework from which we can understand ourselves in relation to the earth.

But human cultures do not snap like bent twigs. As the rise of citizens' juries show, they bend and adapt to change.

What can we in the West do to help those in need through hard times? In the words of Dr. Egziabher, "Make money available to buy food locally and distribute it. If that is not enough, you buy from neighboring areas and distribute. If that is not enough, you get food in kind.... Priorities should be aimed at making that country competent in storing food and making food available in times of need."[53] This will require alternative income-earning activities for people and, ultimately, fundamental changes in the structures of power. That means justice, not charity. The biotechnology industry's appropriation of philanthropic power must be met with the most energetic forms of collective opposition and solidarity the world has ever seen.

GMOs and the Food Crisis in Zambia

MWANANYANDA MBIKUSITA LEWANIKA

THE ZAMBIAN ECONOMY HAS A HISTORY OF OVER-DEPENDENCE ON the mining industry, but increased investments in agriculture during the 1970s attempted to correct this state of affairs by steadily increasing agricultural production. The main difficulty was the management of post-harvest losses due to inadequate storage facilities. In the early 1990s, the introduction of World Bank and International Monetary Fund Structural Adjustment Programs halted government involvement in agricultural production. The government removed agricultural subsidies and stopped procuring inputs such as seeds and fertilizer. This adversely affected small-scale farmers who produced 80 percent of the food, reduced food production, and made it difficult for the government to manage food emergencies.

Zambia experienced a food crisis during the 2001-'02 agricultural season due to unfavorable weather conditions. The crisis was most acute in the Southern Province of Zambia and some parts of the Eastern, Central, Western, and Lusaka Provinces. There was nothing unique about this; in Southern Africa, drought or cyclical phenomena such as El Niño sometimes cause food shortages and even famine.

In response to the food crisis, the World Food Program (WFP) of the United Nations (UN) offered Zambia food aid in the form of corn. The WFP informed the Zambian government that some of the corn was genetically engineered (GE) only after it was already in the country. Three institutions — the National Science and Technology Council, the National Institute for Scientific and Industrial Research, as well as the Soils and Crop Research Branch of the Ministry of Agriculture and Cooperatives — independently advised their respective government ministries against accepting this GE food aid.

The Zambian government advised the WFP not to distribute the GE corn until further notice, and called for a national consultation on whether the country should accept such aid. There were public meetings, interactive radio and television programs, and newspaper articles. Some people contributed to the national consultation on GE food aid by writing letters to the newspaper editors. Even Zambians living outside the country expressed their views on genetic engineering in general, and on the particular issue of GE food aid, in our newspapers.

The national consultation culminated in a national public debate on genetically engineered foods, with Zambian citizens from all walks of life participating. Prominent among the participants were traditional leaders, members of Parliament, representatives of non-governmental organizations, scientists, university lecturers and professors, senior civil servants, and representatives from UN agencies and the donor community, along with ordinary people. Two government ministers attended the debate — the minister of agriculture and co-operatives and the minister of science, technology and vocational training — and the secretary to the cabinet chaired the session.

As a biochemist and negotiator for Zambia around the Cartagena Biosafety Protocol (see Chapter 3), I presented a position paper prepared jointly by three institutions: the National Science and Technology Council; the National Institute for Scientific and Industrial Research; and the Soils and Crop Research Branch of the Ministry of Agriculture and Co-operatives. Participants in the national debate were invited to react to the position paper and add their comments. A representative of traditional leaders said they would support the advice of the Zambian scientists, and an overwhelming majority of participants spoke against accepting the GE food aid. A report of the debate was presented to the government, recommending that the aid should not be accepted.

After further government deliberations, the minister of information and broadcasting announced to the nation and the world that the Zambian government would not accept the aid. The minister made it clear that the decision did not indicate a lack of appreciation for the assistance that was offered. He urged all well wishers to help the country by identifying sources of non-GE food aid.

Like most African nations, Zambia currently has no regulatory system or appropriate infrastructure to cope with the scientific assessments that should accompany the introduction of GE products. There is still great uncertainty about the safety of GE foods for human and animal consumption, as well as for the environment. This led the Zambian government to evoke the Precautionary Principle (see Chapter 2). It was clear that the GE food aid was brought into Zambia without the "advance informed consent" of the Zambian authorities, which is now required under the Biosafety Protocol.

Our health concerns were based on several findings. First, GE foods might contain new toxins or allergens and might increase antibiotic resistance because of the widespread use of antibiotic resistance genes in the creation of GE products. It was noted that millions of Americans consume GE corn, but mostly in processed foods such as corn flakes and taco chips where the transgenic DNA can be broken down during the processing of these products. In contrast, Zambians eat unprocessed corn as their staple food, and usually as their only source of carbohydrates, so the impact would be different. Corn is consumed for breakfast, lunch, supper, and as a snack between meals. Further, the likely recipients of food aid are the most vulnerable members of the society: the old, women, and children, some of whom are immuno-compromised and in poor health.

Our environmental concerns were based on the fear that traditional corn varieties could be genetically contaminated. Since the aid had come in the form of whole grain, some recipients would likely save a portion of it for planting. This could lead to the loss of agricultural diversity in Zambia. We were well aware of the contamination of Mexican native corn due to gene flow from US corn varieties, as reported in the journal *Nature* (See Chapter 1).[1] There were other mitigating factors, such as the worry that Zambian agricultural exports to the European Union (EU), which rejects GMOs, could be adversely affected if any contamination were to occur.

Non-GE corn was already available in some parts of Zambia, in southern Africa, and elsewhere in the world. Our northern region had a surplus of corn, but Zambia lacked the resources to transport it to food-deficient areas. A number of African countries had surpluses of non-GE corn. Even in the United States, only 30 percent of the corn is genetically engineered, and it is possible to segregate non-GE corn from the GE corn.

The Zambian government made its decision not to accept GE food aid in July and August of 2002 and the food crisis was only expected to become critical in March and April of 2003. This gave well wishers enough time to find sources of non-GE food aid. Instead, substantial resources and time were spent trying to convince the government to reverse its decision. Zambian scientists were invited to undertake a fact-finding mission to the US, South Africa, the United Kingdom, the

Netherlands, Norway, and Belgium.

Contrary to comments in the media and arguments by some international non-governmental organizations (NGOs), the government's decision was not made under pressure from NGOs nor the EU. The decision was entirely the result of internal consultations in Zambia, based in part on a scientific assessment of GE foods that called for the use of the Precautionary Principle. The heaviest pressure on the Zambian government to accept GE food aid came from agencies of the UN, especially the WFP, World Health Organization (WHO), and the Food and Agriculture Organization (FAO). These three agencies issued a joint statement to the effect that African countries had no reason to refuse GE food aid, since GE foods are consumed by millions of people globally and allegedly no adverse effects had been observed thus far.

The WFP insisted that it could not source non-GE food aid, and that it could not provide resources to transport corn from surplus areas of Zambia to deficient areas. The WFP insisted that it would not source non-GE corn from within the region because it could obtain corn through open bidding on the international market. The WHO went as far as inviting health ministers from southern Africa to discuss the issue of GE foods.

Senior US officials also pressured Zambia. In his address to the World Summit on Sustainable Development in Johannesburg, US Secretary of State General Colin Powell stated that there was no reason for African countries not to accept GE foods since Americans consume them. US Secretary of Agriculture Anne Veneman blamed anti-biotech forces for scaring Zambians into believing that GE corn would harm them, and the US representative to the FAO was reported to say that those responsible for Zambia's decision should be tried for crimes against humanity.

The Zambian incident had escalated into a full-blown diplomatic row. The US accused "anti-science" Europeans of persuading the Africans into believing that genetically modified foods might be unsafe. In turn, Europeans suggested that the US was cynically trying to shove GE corn they could not sell elsewhere down the throats of starving Africans, while calling it charity. By June of 2003, Agriculture Minister Mundia Sikatana announced that Zambia had produced enough grain to feed all of our people and, by the end of the year, there were sufficient surpluses to export over 50,000 metric tons of corn to neigboring countries.[2]

The Zambian government did not believe that it had made an irresponsible decision by not accepting GE food aid. It was not a question of letting its citizens die of starvation. Early warnings of a food crisis had provided the WFP with enough time to look elsewhere for non-GE food aid. Clearly, there was plenty of non-GE corn in the world's granaries. The real problem was that UN agencies, US officials, and other international players failed to respect Zambia's choice, and then failed to take timely and helpful action.

The Great Trade Robbery:
World Hunger and the Myths of Industrial Agriculture

DEVINDER SHARMA

T WAS THE HEIGHT OF THE GLOBAL CONTROVERSY SURROUNDING genetically engineered food aid to Africa during the late summer of 2002. Hundreds of people in the United States, mostly agricultural scientists, signed an AgBioWorld Foundation petition appealing to the seed multinational giant Aventis CropScience (now part of Bayer) to donate some 3000 tons of experimental genetically engineered (GE) rice to the world's needy rather than destroy it. Behind this call to feed the hungry, however, was a public relations exercise designed to show that biotechnology proponents are concerned about the world's poor.

Aventis had expressed its concern for the hungry in the world, stating that it was "working hard to ensure that US farmers can grow abundant, nutritious crops and we hope that by contributing to that abundance all mankind will prosper." At the same time, the industry-friendly AgBioWorld Foundation conveyed its "disapproval of those who, in the past, have used situations similar to this one to block approved food aid to victims of cyclones, floods and other disasters in order to further their own political [*i.e.*, anti-biotechnology] agendas."[1]

Eradicating global hunger is certainly a noble intention. The US agri-biotech industry and the scientific community had made their stand in defense of a mere 3000 tons of genetically modified rice, the human health risks of which are yet to be ascertained. Yet when some of the appeal's signers were informed that India had a food surplus of 65 million tons, non-genetically modified, and yet a staggering population of some 320 million people went to bed hungry every night, they were not interested. Suddenly, their concern for feeding the hungry evaporated and "humanitarian intentions" vanished into thin air when the fate of their precious GE rice was no longer the issue.

What about the so-called international community? Is it genuinely concerned with the plight of the hungry? During the first World Food Summit (WFS) at Rome in 1996, world leaders "reaffirmed the right of access to safe and nutritious food, consistent with the right to adequate food and fundamental right of everyone to be free from hunger."[2] They deemed it unacceptable that more than 840 million people did not have enough food to meet their basic nutritional needs.

The WFS vowed to feed half of the hungry 840 million by 2015; so, therefore, it would take another 20 years to provide food to the remaining 400 million. They simply postponed this monumental task. By the time 2015 dawns, however, the number of hungry will have multiplied to 1.2 billion. So the world's heads of state were actually admitting their helplessness in tackling hunger and malnutrition. They returned to Rome for the "WFS plus Five" in June 2002, intending to take stock of their progress in reducing hunger over the previous five years. But some 24,000 people were still dying every day from hunger and related diseases. By 2015, more than 122 million people will have succumbed to what is perhaps humankind's greatest shame: hunger in a time of plenty. Politicians never acknowledge that a lack of political will exacerbates hunger and destitution, nor that political power is being exercised to promote technologies and strategies that actually create these imbalances. Instead, they join hands with industrialists and agricultural scientists to chant the *mantra* that genetic engineering will boost food production and solve hunger and malnutrition.

At the same time, free trade and globalization are driving highly subsidized agricultural products from the wealthiest countries into the majority world. Cheap and highly subsidized food imports have already marginalized small farmers, eliminated supplementary livelihoods, and undermined other poverty-coping strategies for millions of agricultural workers in the developing world. Importing food is like importing unemployment, and the world is fast heading toward a situation where developing countries are left with little choice but to remain dependent on the West for their basic food requirements.

Free trade and market domination of food and agriculture are clearly inadequate means for getting food to those who need it most. Market intervention in developing countries is geared toward supporting commercial farmers and export crops, even though projections indicate that the demand for food grains can be met mostly from domestic production. Grain production within the developing world is needed more than imports from the developed world, and food-insecure populations need the income that is generated through employment in agriculture, not just the physical availability of food.

The twin engines of economic growth — the technological revolution and globalization — only widen the existing gap. Biotechnology will push more people into the hunger trap, as public attention and resources are diverted from on-the-ground realities. Simply launching a frontal attack on hunger to ensure that food reaches those who need it most could drastically reduce if not eliminate the problem for 35 to 40 per cent of the world's hungry and malnourished. The world does not have to wait until 2015 to reduce the number of those who are hungry by half.

THE BIOTECH AGE

Biotech proponents insist that the technology of genetic modification holds great promise as the new driving force for sustained agricultural productivity growth in the 21st century. In their view, such growth is crucial if the world is to produce enough food for what is likely to be a stable but much larger world population later in this century. They offer numerous scenarios, projected over different time horizons, so as to create the necessity for biotechnology to step in.[3]

The belief that a biotechnological silver bullet can "solve" hunger, malnutrition, and real poverty has prompted the industry and the development community, as well as the political masters, policy-makers, agricultural scientists, and economists to see an immense potential in "harnessing technology to address specific problems facing poor people."[4] In the bargain, biotechnology comes packaged with stronger intellectual property rights that deprive the developing countries of the ability to undertake research aimed at their own specific needs. Add the usurping of farmers' traditional rights, including the introduction of "Terminator" technology and other similar "genetic use restriction technologies" (GURTs) to prevent seed saving, and the technological dominance of the corporate world is complete.

Even the generally respected United Nations Development Program (UNDP) annual Human Development Report, *Making New Technologies Work for Human Development*, fell into the same trap. On the one hand, the controversial document reported that "technology is created in response to market pressures — not the needs of poor people, who have little purchasing power."[5] On the other hand, it eulogized the virtues of a risky technology that is being imposed on gullible, resource-poor communities of the South in the name of eradicating hunger and poverty.

The report claims that emerging research centers throughout the developing world are already providing hard evidence that cutting-edge science and technology, as biotechnology is routinely described, has the potential to tackle centuries-old problems of poverty. But it does not comprehend that the biggest challenge facing the global community — growing hunger and poverty in the developing countries — requires a social and political commitment rather than a market-driven technological agenda.

Further, the UNDP claimed that "if the developing community turns its back on the explosion of technological innovation in food, it risks marginalizing itself."[6] This in reality is a desperate effort to ensure that corporate interests are not sacrificed on the altar of development. The growing isolation of the GE food industry leads to all kinds of permutations and combinations of misguided policy, including the force-feeding of GE food to the hungry populations in southern Africa.

The reality of hunger and malnutrition is too harsh to even be properly understood from within the cloistered world of development policy. Hunger cannot be ended by producing transgenic crops with genes for beta-carotene, the chemical precursor of vitamin A, or by providing mobile phones to rural communities. Nor can it be eradicated by offering the poor and hungry an "informed choice" of novel GE foods. Even if genetic engineering could reliably increase food production, it would offer no insurance against hunger.

In its effort to bolster the commercial interests of the biotechnology industry, the international community can't see the forest for the trees. In its enthusiasm to promote an expensive technology at the expense of the poor, it overlooks the potential of biotechnology to further deepen the great divide between the haves and have-nots. No technological fix can help bridge this monumental gap. Hunger results not only from the inability of the poor to access food, it is also the result of global policies that further marginalize the poverty-stricken, cumulatively adding to the problem of hunger.

While political leaders and the development community stall in their commitment to halve the number of the world's hungry, the scientific community has found an easy escape route. At nearly all the genetic engineering laboratories, whether in the North or South, research is focused on transgenic crops that can produce edible vaccines as well as Vitamin A, iron, and other micronutrients to address malnutrition. But unless hunger is systematically eradicated, the "hidden hunger" of particular nutrient deficiencies cannot be prevented.

In India, which is technically self-sufficient in food grain production, reports of hunger and starvation regularly pour in from the
notoriously poverty-stricken Kalahandi region and more recently from
Kashipur in Orissa on the eastern coast.[7] With a population of 20 million, the region suffers the pangs of hunger and malnutrition despite
the absence of visible ecological devastation. Kalahandi is a fertile
area, a traditional food basket; at the time of the 1943 Bengal Famine,
it even came to the rescue of the afflicted regions.

The problem is certainly not lack of production. The Kalahandi is
the biggest contributor of surplus rice to Orissa's state food reserves.
Between 1996 and 2001, it provided an average of 50,000 tons of surplus rice per year. People do not die of starvation and hunger because
there is not enough food, rather because they cannot afford to buy
the food they produce. Biotechnology cannot help to ensure that food
comes within reach of these poorest of the poor.

Biotechnology is said to be the only or the best "tool of choice"
for marginal ecological zones left behind by the Green Revolution
and still home to more than half of the world's poorest people. These
people depend on agriculture and livestock, even while living in often
harsh and inhospitable areas. The way technology is being applied
and blindly promoted, however, makes it obvious that the underlying
interests are not altruistic.

Risky technology is being pushed into use before farmers, consumers, and local policy makers can assess its negative consequences.
Even the judiciary is being co-opted. In India, for example, a visiting
delegation of 10 US judges and scientists met with A.S. Anand, then
chief justice of India, to impress upon him and the entire judicial community the benefits of biotechnology. Reports in *The Hindu* quoted
Dr. Franklin M. Zweig, president of the Einstein Institute for Science,
Health and the Courts in the US, who spoke in favor of genetic engineering at the 88th session of the Indian Science Congress in New
Delhi in January 2001.[8] Under pointed questioning, Dr. Zweig denied
that the two-hour meeting with the chief justice was intended to influence the judiciary, but rather aimed at "educating" the judges about
basic principles of public information for use by the courts and court
systems.

The delegation also invited India's chief justice to the US and
offered to hold workshops for the judges of the Supreme Court and

the High Court to "educate" them about transgenics and safety proto-
cols in biotechnology research. The delegation, which included some
Indian-born US scientists, proposed that nations work together to set
"ethical guidelines" on genetic engineering. Working groups of the
Einstein Institute in the Philippines, South Africa, Israel, Italy, the
UK, Netherlands, and Canada have made similar attempts.

GOLDEN BLUFF

In a frantic effort to repair its damaged credibility, the biotechnology industry is set to unleash a "secret weapon" on millions of unsus-
pecting, destitute smallholders in the developing world. This particular
weapon, "golden rice," is also an ecological and health hazard. At best
it can provide a miniscule amount of micronutrients; it cannot address
the nutritional needs of small farmers and the poverty-stricken masses.
Their remaining requirements will have to be met through other
nutritional sources. In India, for instance, rice is invariably consumed
with a combination of *pulses* (legumes), such as lentils and peas, which
provide the essential proteins and vitamins the human body needs. The
situation is the same in other developing countries.

Syngenta's Dr. Adrian Dubock recently claimed that "the levels of
expression of pro-vitamin A that the inventors were aiming at, and
have achieved, are sufficient to provide the minimum level of pro-vitamin A to prevent the development of irreversible blindness affecting
500,000 children annually, and to significantly alleviate Vitamin A deficiency affecting 124 million children in 26 countries."[9] He also warned
that for each month the introduction of "golden rice" into the market
is delayed, 50,000 children would go blind. However, a simple calcu-
lation based on recommended daily allowance (RDA) figures shows
an adult would have to eat at least 12 times the normal intake of 300
grams of rice to get the daily recommended amount of pro-vitamin A
from "golden rice." Pro-vitamin A is a biochemical precursor that is
metabolized to vitamin A only in the presence of sufficient body fat.

Over the centuries, societies have evolved and perfected dietary
systems that provide the nutrient balance humans require. Yet, micronutrient deficiency in human food is nothing new. One perplexing
question here is: Who decided that vitamin A is the most essential
micronutrient to be incorporated in rice? Why not vitamin B complex? After all, several hundred million people in India alone suffer
from malnutrition, compared to only half a million people worldwide

who go blind from vitamin A deficiency. In India alone, some 12 million people suffer from vitamin A-deficiency, but the number lacking adequate vitamin B complex is several times higher.

Under an Indo-Swiss collaboration, "golden rice" technology will be made available to the Indian Council for Agricultural Research (ICAR) and the Indian Department of Biotechnology. The project, funded to the tune of US $ 2.6 million over seven years, aims to engineer pro-vitamin A genes into local varieties of rice. In reality, ICAR's tryst with golden rice is a case of blind experimentation, and a misguided attempt to regain its lost pride in agricultural research. Suffering from a credibility crisis in the absence of any significant breakthrough after the initial phase of the Green Revolution, ICAR is eager to divert attention from more pressing problems confronting rural society.

A majority of the acutely malnourished people whom the proponents of "golden rice" claim they want to help cannot afford to buy rice from the market. If these poor people cannot buy ordinary rice, how will they pay for "golden" rice? The question has been conveniently overlooked. If these hungry millions could meet their daily rice requirement, there would be no malnutrition in the first place. The problem cannot be solved by providing nutritional supplements through GE rice. The answer lies in policy changes that force governments to ensure sufficient food for all.

PRODUCE AND PERISH

At the height of the rice paddy harvesting season in September 2000, hundreds of thousands of farmers in northwest India's frontline agricultural states of Punjab, Haryana, and western Uttar Pradesh waited patiently for buyers until government agencies were finally forced to purchase their excess stocks. For three weeks, the farmers sat patiently over heaps of rice in the grain markets. At least 100 farmers, unable to bear the economic burden, chose to commit suicide by drinking pesticides. In Andhra Pradesh, in south India, there were no buyers for a five-million-ton surplus. Even in the poverty-stricken belt of Bihar and Orissa, in north-central India, farmers waited in vain.

Farmer suicides reflect the breakdown of institutional safety nets that have cushioned the impact of agrarian crises in the past. Agricultural scientists want farmers to go on producing more food grains, keeping the food deficit projections for 2015 in mind. Yet the state government of Andhra Pradesh has publicly asked farmers not to produce more paddy rice. In Punjab, the citadel of the Green Revolution, farmers are being asked to shift from staple foods like wheat and rice to cash crops such as tomatoes and cut flowers. There appears to be something terribly wrong with this formula. Scientists, industry, and government planners blindly support biotechnological breakthroughs in the name of feeding the world, while political masters in the developing world are asking farmers not to produce more.

What lies ahead is frightening. Committed to the World Trade Organization, the Indian government has begun to remove trade barriers, and we have become an agricultural dumping ground. Consider the case of edible oils. The government has reduced customs duties, opening up the country to increasing imports. Starting from a position of strength, India has turned into a major importer of edible oils over the past decade, heedless of the impacts such imports have on the livelihoods and security of millions of oilseed farmers. With the "blue revolution" already sucked dry by the aquaculture industry, a "yellow revolution" in oilseeds turning colorless, and the "white revolution" under attack from the private dairy industry and cheap imported milk products, all eyes are now set on dismantling the ever-diminishing gains of the much-acclaimed Green Revolution.

Consider the following: At the time of its independence in 1947, India had about five million farms. By the early 1980s, the number had risen to about 90 million, and the most recent estimate is that there are now some 110 million farms in the country. Every fourth farmer in the world today is Indian, and nearly half the country's land is being utilized for crop production. Already, India's population has crossed the one billion mark. India also has 20 percent of the world's livestock. Still, more than 320 million people, mostly women and children, suffer from chronic hunger at a time when grain silos are bursting at the seams.

In 2000, India had a record food surplus of 44 million tons. By 2002, the surplus had grown to 65 million tons, not due to excess production, but because more and more people were unable to buy the grain that lay stockpiled. Ironically, while the Indian government is asking its farmers to diversify from rice and wheat to cash crops, the National Agricultural Policy projects that an annual four percent growth rate in food grains will be needed to meet the country's requirements in the years to come. Regardless of what the agricultural scientists and policy makers say, the government is slowly but steadily dismantling the procurement system and pricing policies that served as effective instruments to usher in food self-sufficiency following the Green Revolution.

The choice, therefore, is limited. The only viable path towards sustaining the natural resource base to satisfy the demands of the growing population for food and other agricultural commodities lies in enhancing the potential of domestic agriculture. A correct mix of policies, coupled with strategies that restore the status of Indian agriculture, can help provide an answer to the food crisis. This requires location-specific technologies and production packages that meet the aspirations of the majority of the farmers who own fewer than 2 hectares on average. The scenario and the solutions are certainly no different for a majority of the developing countries.

The paradox of plenty is not confined to India. Until recently, Pakistan, Bangladesh, and even Indonesia overflowed with surplus grains. In Pakistan, farmers have burned harvested grains lying in markets because the prices were too low. In Indonesia, farmers wait endlessly to sell rice while the government imports it from Vietnam. Strangely, those who swear in the name of hunger seem not to know that farmers in India and across Asia are waiting endlessly for someone to buy the grains they harvest.

Should farmers therefore continue to "produce and perish," despite agricultural scientists' repeated assertion that the world will need to produce more food in 2015? Should farmers be driven to commit suicide for producing more? Should the poor and hungry wait for the year 2015 to get a morsel of food while the contents of overflowing grain silos are devoured by insect pests and rodents in front of their eyes?

Who will take on the biggest challenge of all times: the elimination of hunger, which forms the root cause of real poverty? By leaving the monumental task of feeding the hungry to the market — especially at a time when food lies rotting — politicians, industry, and the scientific community are merely abdicating their responsibility to feed the world now. No "shock therapy" is needed to convince the world that a crisis will unfold by the year 2015 or perhaps 2020.

The only solution is to implement national and international policies that actually encourage self-sufficiency and by sustainable means. Leave the rest to farmers. They have done it in the past and they have the capacity to do it in the future.

Endnotes

INTRODUCTION

1 The terminology of "neoliberalism" evokes the original meaning of "liberalism" in Western discourse: the view that a self-regulating "free market" should be the sole arbiter of social and economic policies. On the origins of the "free market" myth in Europe, see Karl Polanyi, *The Great Transformation: The Political and Economic Origins of Our Time*, Boston: Beacon Press: 1957.

2 John Losey, et al., "Transgenic pollen harms monarch larvae," *Nature* Vol. 399, 1999, p. 6733.

3 For a full discussion of this development, its history and implications, see Brian Tokar, *Earth for Sale: Reclaiming Ecology in the Age of Corporate Greenwash*, Boston: South End Press, 1997.

4 These events have featured several of the largest-ever protests against genetic engineering in North America. See www.biodev.org.

5 Jim Thomas, "Princes, Aliens, Superheroes and Snowballs: The Playful World of the UK Genetic Resistance," in Brian Tokar, ed., *Redesigning Life? The Worldwide Challenge to Genetic Engineering*, London: Zed Books, 2001, pp. 337-350.

6 José Bové and Francois Dufour, *The World is Not for Sale: Farmers Against Junk Food*, London: Verso Books, 2001.

7 Martin Khor, "500,000 Indian farmers rally against GATT and patenting of seeds," Third World Resurgence, Number 39, November 1993; Vandana Shiva, "Seed Satyagraha: A movement for farmers' rights and freedoms in a world of intellectual property rights, globalized agriculture and biotechnology," in Brian Tokar, ed., *Redesigning Life?*, pp. 351-360.

8 Vandana Shiva, et al., *The Seed Keepers*, New Delhi: Navdanya, 1995.

9 See www.mstbrazil.org.

10 "Brazilian farmers storm Monsanto, uproot plants," Reuters, January 26, 2001.

11 The full text is available at http://www.biodev.org/archives/000170.php.

12 See http://www.biodev.org/archives/000185.php and http://www.bite-back.org.

13 Brian Tokar and Doyle Canning, "Countering Biotech and 'Free Trade' in Sacramento," Z *Magazine*, September 2003, pp. 20-25.

14 See Brian Tokar, "Monsanto: A Profile of Corporate Arrogance," in Edward Goldsmith and Jerry Mander, eds., *The Case Against the Global Economy*, London: Earthscan, 2001, and "Monsanto: A Checkered History," *The Ecologist*, September/October 1998.

15 Peter Shinkle, "Monsanto reaps some anger with hard line on reusing seed," *St. Louis Post-Dispatch*, May 12, 2003.

16 Stanley W.B. Ewen and Arpad Pusztai, "Effect of diets containing genetically modified potatoes expressing Galanthus nivalis lectin on rat small intestine," The Lancet, Vol. 354, No. 9187, October 16, 1999, pp. 1353-54. For a description of Dr. Pusztai's experiments and his subsequent blacklisting by the UK scientific establishment, see Jeffrey Smith, *Seeds of Deception: Exposing*

Industry & Government Lies About the Safety of the Genetically Engineered Foods You're Eating, Fairfield, IA: Yes Books, 2003.

17 See, for example, A. R. Zangerl, *et al.*, "Effects of exposure to event 176 Bacillus thuringiensis corn pollen on monarch and black swallowtail caterpillars under field conditions"; Diane E. Stanley-Horn, *et al.*, "Assessing the impact of Cry1Ab-expressing corn pollen on monarch butterfly larvae in field studies"; Karen S. Oberhauser, *et al.*, "Temporal and spatial overlap between monarch larvae and corn pollen," all published in the *Proceedings of the National Academy of Sciences*, Vol. 98, No. 21, October 9, 2001; also John J. Obrycki, *et al.*, "Transgenic Insecticidal Corn: Beyond Insecticidal Toxicity to Ecological Complexity, *BioScience*, Vol. 51, No. 5, May 2001, pp. 353-361.

18 For a comprehensive overview, see Ricarda Steinbrecher, "Ecological Consequences of Genetic Engineering," in Brian Tokar, ed., *Redesigning Life?* pp. 75-102.

19 Margaret Mellon and Jane Rissler, *Gone to Seed: Transgenic Contaminants in the Traditional Seed Supply*, Washington, DC: Union of Concerned Scientists, 2004.

20 The experiment required 242 human eggs, "donated" by 16 women, presumably under the influence of drugs designed to massively increase egg production (superovulation). See, for example, Gina Kolata, "Cloning Creates Human Embryos," *New York Times*, February 12, 2004.

21 Bill Lambrecht, "Biotech rivals team up in effort to sell altered food," *St. Louis Post Dispatch*, April 4, 2000.

22 See, for example, Bill Lambrecht, "Monsanto battles effort to require labeling of genetically modified food," *St. Louis Post-Dispatch*, September 19, 2002; "CF&M Heads Anti-Label PR Push for Biotechs," *O'Dwyer's PR Daily*, September 30, 2002, at http://www.odwyerpr.com/archived_stories_2002/september/0930gen.htm; Norfolk (UK) Genetic Information Network, "Biotech money-in-politics," *Weekly Watch*, No. 34, at http://www.gmwatch.org; Ken Garcia," Mendocino sows seeds of dissent," *San Francisco Chronicle*, March 8, 2004.

23 Michael Pollan, "The Great Yellow Hype," *New York Times Magazine*, March 4, 2001; Vandana Shiva, "Genetically Engineered 'Vitamin A Rice': A Blind Approach to Blindness Prevention," in Brian Tokar, ed., *Redesigning Life?* pp. 40-43.

24 Vandana Shiva, "Biotech Companies As Bioterrorists," *ZNet* Commentary, November 6, 2001, at http://www.zmag.org/sustainers/content/2001-11/06shiva.cfm.

25 ETC Group, "Maize Rage in Mexico: GM maize contamination in Mexico, 2 years later," *ETC Group Genotypes*, October 10, 2003, at http://www.etcgroup.org.

26 Jules Pretty and Rachel Hine, *Reducing Food Poverty with Sustainable Agriculture: A Summary of New Evidence*, University of Essex, UK, February 2001; Fred Pearce, "An Ordinary Miracle," *New Scientist*, Vol. 169, No. 2276, February 3, 2001, p. 16; Jules Pretty, "Against the Grain: Could we feed the world without causing further environmental damage?" *The Guardian*, January

17, 2001.

27 Vandana Shiva, *Staying Alive: Women, Ecology and Development*, London: Zed Books, 1988, p. 103.

28 Vandana Shiva, *The Violence of the Green Revolution: Third World Agriculture, Ecology and Politics*, London and New Jersey: Zed Books, 1991, p. 15.

29 "Food as a Political Weapon" (interview with Devinder Sharma), *Acres USA*, March 2004, at http://www.acresusa.com/magazines/archives/0304SharmaInterview.htm.

30 Les Levidow and Joyce Tait, "The Greening of Biotechnology: From GMOs to Environment-Friendly Products," *Occasional Paper 21 of the Open University Technology Policy Group*, Milton Keynes, UK, August 1990.

CHAPTER 1

1 Marc Kaufman, "The biotech corn debate grows hot in Mexico," *Washington Post*, March 25, 2002.

2 Michael Hansen, "Genetic engineering is not an extension of conventional plant breeding," Consumers Union, 1998, p. 2, available at http://www.consumer-sunion.org/food/widecpi200.htm.

3 Ricarda Steinbrecher, "Ecological consequences of genetic engineering," in Brian Tokar, ed., *Redesigning Life? The Worldwide Challenge to Genetic Engineering*, London: Zed Books, 2001, p. 83.

4 Martha L. Crouch, "From Golden Rice to Terminator Technology: Agricultural biotechnology will not feed the world or save the environment," In Brian Tokar, ed., *Redesigning Life?*, p. 29.

5 R. Steinbrecher, "Ecological consequences of genetic engineering," *op. cit.*, pp. 84-86.

6 *Ibid*, pp. 84-85, and references therein.

7 Matthew Gledhill and Peter McGrath, "Call for a spin doctor," *New Scientist*, November 1, 1997.

8 Claire Gilbert, "Honeybees and Genetically Engineered Canola Oil," October 1997, available at http://www.pmac.net/canola.htm.

9 R. Steinbrecher, "Ecological consequences of genetic engineering," *op. cit.*, p. 84.

10 D. Saxena, S. Flores and G. Stotzky, "Insecticidal toxin in root exudates from Bt corn," *Nature*, Vol. 402, 1999.

11 Tom Block, "Pseudopregnancies puzzle swine producer," Iowa Farm Bureau, April 29, 2002.

12 Network of Concerned Farmers, "Farmers seek action regarding GE corn batch causing breeding problems," January 23, 2003, p. 1, Available at http://www.non-gm-farmers.com.

13 Andrew Pollack, "Widely Used Crop Herbicide Is Losing Weed Resistance", *New York Times*, January 14, 2003.

14 *Ibid.*

15 Charles M. Benbrook, *Impacts of Genetically Engineered Crops on Pesticide Use in the United States: The First Eight Years*, BioTech InfoNet Technical Paper Number 6, November 2003, available from http://www.biotech-info.net.

16 Joseph Mendelson, "Roundup: The world's biggest selling herbicide," *The*

Ecologist, Vol. 28, No. 5, September/October 1998, pp. 270-275.

17 *Ibid.,* p. 272.

18 *Ibid.*

19 Caroline Cox, "Glyphosate," *Journal of Pesticide Reform,* Vol. 11, No. 2, Summer 1991.

20 Lennart Hardell and Miikael Eriksson, "A Case-Control Study of Non-Hodgkin Lymphoma and Exposure to Pesticides," *Cancer,* Vol. 85, No. 6, March 15, 1999, See also *Rachel's Environment and Health News,* Number 751, September 5, 2002, at http://www.rachel.org.

21 Joseph Mendelson, "Roundup: The world's biggest selling herbicide," *op. cit.*

22 *Ibid.,* p. 274.

23 *Ibid.,* p. 272.

24 Martha Crouch, "From Golden Rice to Terminator Technology," *op. cit.*

25 Angélica Enciso, "Experts alert about the release of this corn variety into the market in the near future: Contraceptive corn, a danger: it could cause human sterility," *La Jornada,* January 25, 2002.

26 Philip Brasher, "Biotech company to isolate plants," *Des Moines Register,* November 16, 2002, at http://www.neRAGE.org/stories.php?story=02/11/16/0 939383.

27 Larry Bohlen, "Biopharming: 'A new potential form of GMO contamina-tion,'" in *GMO Contamination around the World,* Amsterdam: Friends of the Earth International, 2002, p. 20; See also Brian Tokar, *Biohazards: The Next Generation? Companies Are Genetically Engineering Crop Plants that Manufacture Industrial and Pharmaceutical Proteins,* Edmonds, WA.: Edmonds Institute, 2001, available at http://www.neRAGE.org/stories.php?sto ry=02/03/20/1083111.

28 Testimony by Dr. Robert Drotman of the Frito Lay Company to the Texas House Agriculture and Livestock Committee, April 10, 2003, forwarded by Bill Freese of Friends of the Earth.

29 Philip Brasher, "Biotech company to isolate plants," *op. cit.*

30 *Ibid.*

31 Center for Food Safety, "The hidden health hazards of genetically engineered foods," *Center for Food Safety Review,* Spring 1999.

32 Jennifer Ferrara and Michael Dorsey, "Genetically Engineered Foods: A mine-field of safety hazards," In Brian Tokar ed., *Redesigning Life?,* p. 63.

33 Friends of the Earth, *GMO Contamination around the World,* Amsterdam: Friends of the Earth International, 2002, p. 12.

34 *Ibid.* p . 10.

35 John Ross, "Tinkering with the tortilla: Genetic engineering threatens Mexico's corn culture," *Sierra Magazine,* Vol. 86, No. 5, September/October 2001, p. 21.

36 Anthony Shadid, "Tests for genetic corn spur concerns amounts of disputed vari-ety found in 10% of US grain inspections," *Boston Globe,* May 3, 2001.

37 Friends of the Earth, *GMO Contamination around the World, op. cit.,* p. 11.

38 *Ibid.,* p. 14.

39 *Ibid.,* p. 14.

40 *Ibid.*, p. 19.

41 Paul Jacobs, "Banished biotech corn not gone yet," *San Jose Mercury News,* December 1, 2003.

42 Jennifer Ferrara and Michael Dorsey, "Genetically Engineered Foods: A minefield of safety hazards," *op. cit.*, p. 58.

43 Stanley W.B. Ewen and Arpad Pusztai, " Effect of diets containing genetically modified potatoes expressing *Galanthus nivalis* lectin on rat small intestine," *Lancet,* Vol. 354, No. 9187, October 16, 1999, pp. 1353-4.

44 Jennifer Ferrara and Michael Dorsey, "Genetically Engineered foods: A minefield of safety hazards," *op. cit.* p. 58. For an engaging account of Dr. Pusztai's work, see Jeffrey Smith, *Seeds of Deception: Exposing Industry and Government Lies About the Safety of the Genetically Engineered Foods You're Eating,* Fairfield, Iowa: Yes Books, 2003, pp. 5-32.

45 Peter Shinkle, "Monsanto reaps some anger with hard line on reusing seeds," *St. Louis Post-Dispatch,* May 12, 2003.

46 Percy Schmeiser, lecture at the University of Vermont, Burlington, Vermont, February 19, 2003.

47 Roberto González, *Zapotec Science: Farming and Food in the Northern Sierra of Oaxaca,* Austin: University of Texas Press, 2001, p. 120.

48 James Boyce, "The environment and security: The maize connection," *Political Environments,* Vol. 5, September 30, 1997, p. 22.

49 Ignacio Chapela and David Quist, "Transgenic DNA introgressed into traditional maize landraces in Oaxaca, Mexico," *Nature,* Vol. 414, November 2001, pp. 541-543.

50 Paul Brown, "Mexico's vital gene reservoir polluted by modified maize," *The Guardian,* April 19, 2002,

51 *Ibid.*

52 "Contamination by genetically modified maize in Mexico much worse than feared," report released jointly by indigenous and farming communities from Oaxaca, Puebla, Chihuahua, and Veracruz: CENAMI, CECCAM, ETC Group, CASIFOP, UNOSJO, AJAGI, October 9, 2003, available at http://www.etcgroup.org.

53 *Ibid.*

54 John Nagel, "Mexico to lift de facto moratorium on GM corn experiments, official says," *International Environment Reporter,* Vol. 26, No. 21, October 8, 2003, p. 1012.

55 Héctor Magallón, Greenpeace Mexico, interviewed February 21, 2002.

56 Marc Kaufman, "The biotech corn debate grows hot in Mexico," *Washington Post,* March 25, 2002.

57 Chakravarthi Raghavan, "Mexico: NAFTA corn liberalization fails farmers, environment," *Third World Network* : Geneva, October 2002, available at http://www.twnside.org.sg/title/Mexico.htm.

58 *El Campo No Aguanta Más,* flyer from farmer protest in Mexico City, January 31, 2003.

59 Mark Stevenson, "Mexicans angered by spread of corn," Associated Press, December 29, 2001.

60 Philip Brasher, "Biotech crops spread in the US," *Burlington Free Press*, March 29, 2002.

61 Marc Kaufman, "The biotech corn debate grows hot in Mexico," *Washington Post*, March 25, 2002.

62 Manuel Mérida, Unit Boss at Diconsa headquarters in Oaxaca City, interviewed January 6, 2002, in Oaxaca City, Oaxaca, Mexico.

63 CONABIO and INE (Two agencies of Mexican government). Report presented at the "In the Defense of Corn Conference" in Mexico City January 2002.

64 Interview January 12, 2002 in Ixtlán, Oaxaca, Mexico.

65 Interview January 11, 2002 in Guelatoa, Oaxaca, Mexico.

66 S'ra DeSantis, survey for thesis for the University of Vermont: 29 *Campesino* families from five communities. Interviews from January 4, 2002 to January 13, 2002 in Ixtlán, Guelatoa, Calpulalpan, Yavesia, and San Juan, in the Sierra Juárez of Oaxaca, Mexico.

67 Tina Rosenberg, "Why Mexico's small farmers go hungry," *New York Times*, March 3, 2003.

68 Umberto Rosales, engineer with SAGARPA, interviewed January 6, 2002, in Oaxaca City, Oaxaca, Mexico.

69 *El Campo No Aguanta Más*, flyer from farmer protest in Mexico City, January 31, 2003.

70 *Ibid.*

71 Doreen Hemlock and Joseph Mann, "Compromise draws business groups' criticism," *Sun Sentinel*, November 21, 2003, p. A23; Tere Figueras, S. Olkon, and M. Merzer, "Agreement ends summit early," *The Miami Herald*, November 21, 2003.

72 Walden Bello, "Original FTAA draft scrapped: People pour into Miami to protest FTAA," November 22, 2003, available at www.commondreams.org.

73 Warren Vieth, "US reaches Central American trade deal," *Los Angeles Times*, December 18, 2003.

74 Edward Alden, "US agrees trade deal with Central American nations," *Financial Times*, Washington DC, December 18, 2003.

75 Shiri Pasternak, "War by Other Means: Geopolitics and Pop-Tarts in the New Agrarian Order," *Food Not Lawns, Volume II*, 2002, p. 15.

76 Friends of the Earth, *op. cit.*, 2002, p. 17.

77 Charles Warpehoski, "Introduction to the Plan Puebla Panama," in *Plan Puebla Panama: Battle over the Future of Southern Mexico and Central America*, Network Opposed to the Plan Puebla Panama, 2002, pp. 5-7.

78 La Coordinadora Regional de Los Altos de Chiapas, "Maquiladoras: "The march of sweatshops toward the south," in *Plan Puebla Panama: Battle over the Future of Southern Mexico and Central America, ibid.*, p. 19.

79 Personal communication, February 4, 2003.

80 Personal communication, February 27, 2003.

CHAPTER 2
1 Renato Ruggiero, Address to UNCTAD Trade and Development Board, London, October 8, 1996, available at http://www.unicc.org/unctad/en/special/

TB43Pro5.htm.

2 Bernard Hoekman and Michel Kostecki, *The Political Economy of the World Trading System: From GATT to WTO*, Oxford University Press, 1995, p. 24.

3 Aroha Te Pareake Mead, "Cultural and Intellectual Property Rights of Indigenous Peoples of the Pacific", workshop presentation, Inaugural Indigenous Peoples of the Pacific workshop on the UN Draft Declaration on the Rights of the Indigenous Peoples, September 2-6, 1996, Suva, Fiji.

4 Helena Paul and Ricarda Steinbrecher with Devlin Kuyek and Lucy Michaels, *Hungry Corporations: Transnational Biotech Companies Colonise the Food Chain*, London, Zed Books, 2003, p.149. See also http://www.edmonds-institute.org/door.html.

5 "Parallel importing" allows retailers, wholesalers, governments, and other parties to obtain goods subject to intellectual property rights directly from licensed or authorized overseas sources, rather than dealing with local suppliers, licensees, or agents. It allows the buyer to shop around for the lowest world price, and thus enables developing countries to get lower prices for consumers for pharmaceuticals and other goods. Although the WTO TRIPs does not prohibit parallel imports, many business lobbies like PhRMA and the US Trade Representative's office have aggressively opposed this practice, claiming that it will undermine intellectual property rights, and threatening countries that permit the practice.

6 "World Patents for Global Domination?" in *Seedling* (GRAIN publications), October 2003, pp. 12-16.

7 George Monbiot, *Captive State: the Corporate Takeover of Britain*, London: Macmillan, 2000, p. 253, quoting from *Farm Journal*, cited by the Rural Advancement Foundation International, September 1996: *The Life Industry*, now at http://www.etcgroup.org/article.asp?newsid=198.

8 In Steven Benowitz, "Technology Motivating Industry", *The Scientist*, Vol. 10, No. 3, February 5, 1996.

9 Hope Shand and Martin Teitel, *The Ownership of Life: When Patents and Values Clash*, Institute for Agriculture and Trade Policy, June 1997 and at http://www.sustain.org/biotech/library/admin/uploadedfiles/Ownership_of_Life_When_Patents_and_Values_Clas.htm.

10 Cited in *The Patenting of Human Genetic Material*, RAFI Communique, January/February 1994.

11 "Taking forward the review of Article 27.3(b) the TRIPs Agreement," communication from the Africa Group, WTO, IP/C/W/404, June 26, 2003 at http://docsonline.wto.org.

12 WTO website, http://www.wto.org/english/thewto_e/minist_e/ min01_e/ mindecl_trips_e.htm.

13 "Swiss NGOs Criticise WTO Drug Agreement", *Swissinfo*, August 31, 2003 at http://www.nzz.ch/2003/08/31/english/page-synd4182839.html.

14 Nick Mathiason, "WTO Drugs Deal 'Not Viable'," *The Observer*, August 31, 2003.

15 Agreement On Agriculture, WTO website, http://www.wto.org/english/docs_e/ legal_e/14-ag_01_e.htm.

16 "Poorer Countries Say WTO Farm Accords Help Rich," Reuters, June 28, 2000.

17 Via Campesina Seattle Declaration, Seattle, December 3, 1999.

18 See, for example, "Sri Lanka's GM food ban delayed indefinitely," Reuters, September 4, 2001.

19 See the US Trade Representative's website, www.ustr.gov.

20 Yasser Sobhi, "In a jam over GM foods," *Al-Ahram Weekly Online*, Egypt, July 10-16, 2003, at http://weekly.ahram.org.eg/2003/646/ec1.htm.

21 See the Council's website, http://www.agritrade.org.

22 "U.S. and Cooperating Countries File WTO Case Against EU Moratorium on Biotech Foods and Crops," USTR Press Release, May 13, 2003, http://www.ustr.gov/releases/2003/05/03-31.htm.

23 See, for example, Ashok B. Sharma, "Third World Must Actively participate in Global Standards Bodies," *Financial Express*, India, November 24, 2003, at http://www.financialexpress.com/fe_full_story.php?content_id=46896.

24 European Union Online, "EU complies with WTO ruling on Hormone beef and calls on USA and Canada to lift trade sanctions," Brussels, October 15, 2003, at http://europa.eu.int/comm/trade/issues/respectrules/dispute/pr151003_en.htm.

25 http://www.codexalimentarius.net.

26 Council for International Business website, http://www.uscib.org/index.asp?documentID=810.

27 See Chapter 3, No. 26.

CHAPTER 3

1 For an up-to-date overview of World Bank policies and practices, see "More World, Less Bank," a special issue of the *New Internationalist*, Number 365, March 2004.

2 Sustainable development was introduced to a non-specialist public with the publication of the World Commission on Environment and Development's landmark report, *Our Common Future*, New York: Oxford University Press, 1988.

3 See, for example, Jim McNeill, *et al.*, *Beyond Interdependence: The Meshing of the World's Economy and the Earth's Ecology*, New York: Oxford University Press, 1991.

4 Kevin M. Cleaver (director, World Bank Agriculture and Rural Development Department), letter to the Pesticide Action Network North America Regional Center, July 30, 2002.

5 Ichiyo Muto, "For an Alliance of Hope," in Jeremy Brecher, *et. al.*, eds., *Global Visions: Beyond the New World Order*, Boston: South End Press, 1993, p. 148.

6 See, for example, "Investing in Destruction — The World Bank and Biodiversity," Barcelona: GRAIN, November 1996, available at http://www.grain.org/publications/bio8-en.cfm.

7 See Vandana Shiva, *Staying Alive: Women, Ecology and Development*, London: Zed Books, 1988, esp. Chapter 4, *The Violence of the Green Revolution: Third World Agriculture, Ecology and Politics*, London and New Jersey: Zed Books, 1991.

8 Marcia Ishii-Eiteman, *et al.*, "Trouble on the Farm: World Bank projects

undermine sustainable agriculture," Pesticide Action Network North America, 2000, at http://www.panna.org.

9 Christine Lee, "All Pain, No Gain: How Structural Adjustment Hurts Farmers and the Environment," *Global Pesticide Campaigner*, Vol. 11, No. 1, April 2001.

10 Eija Pehu, "Biosafety Capacity Building: A World Bank Perspective," in M. A. Mclean, *et al.*, eds., *A Framework for Biosafety Implementation: Report of a Meeting*," ISNAR (International Service for National Agricultural Research) Biotechnology Service, February 2003, pp. 27-30.

11 *Ibid.*, p. 28.

12 *Ibid.*

13 "Providing Proteins to the Poor: Genetically Engineered Potatoes vs. Amaranth and Pulses," New Delhi: Research Foundation for Science, Technology and Ecology, January 9, 2003, archived at http://www.gene.ch/genet.html.

14 Aaron deGrassi, *Genetically Modified Crops and Sustainable Poverty Alleviation in Sub-Saharan Africa: An Assessment of Current Evidence*, Third World Network/Africa, June 2003, pp. 6-10.

15 Gatonye Gathura, "GM technology fails local potatoes," Nairobi, Kenya: *The Daily Nation*, January 29, 2004, at http://www.nationaudio.com/News/ DailyNation/Supplements/horizon/current/story290120041.htm

16 Aaron deGrassi, *Genetically Modified Crops and Sustainable Poverty Alleviation in Sub-Saharan Africa, op.cit.*, p. 8.

17 *Agricultural Biotechnology: The Next "Green Revolution"?* World Bank Technical Paper Number 133, January 1991. As of this report, the Bank had already committed $93 million to biotechnology research efforts in Indonesia, Sri Lanka, Brazil, Africa, and central Europe (p. 33).

18 Genetic Resources Action International (GRAIN), "ISAAA in Asia: Promoting corporate profits in the name of the poor," Barcelona: GRAIN, October 2000, at http://www.grain.org/publications/isaaa-en.cfm.

19 Henry W. Kendall, *et al.*, *Bioengineering of Crops: Report of the World Bank Panel on Transgenic Crops*, Washington, DC: International Bank for Reconstruction and Development/World Bank, 1997.

20 See, for example, Stas Burgiel and Lynn Wagner, "Workshop on Agricultural Biotechnology and Rural Development Priorities for the World Bank," *Sustainable Developments* Vol. 26, No. 1, June 1999, at http://www.iisd.ca/ download/asc/sd/sdvol26no1e.txt.

21 Marcia Ishii-Eiteman, "Pesticides and Pest Management," in *Marketing the Earth: The World Bank and Sustainable Development*, Washington, DC: Friends of the Earth and Halifax, Nova Scotia: The Halifax Initiative, 2002, pp. 27-32.

22 *Ibid.*, p. 30.

23 Details and individual participant profiles are at http://www.staffexchange.org.

24 *Ibid.*

25 Convention on Biological Diversity website, http://www.biodiv.org/biosafety/ ratification.asp.

26 The Precautionary Principle has become a central concept in international environmental policy since the 1990s, with greatest acceptance in the countries

of the European Union. A major international conference on the principle in 1998 defined it this way: "When an activity raises threats of harm to the environment or human health, precautionary measures should be taken even if some cause and effect relationships are not fully established scientifically." ("Precautionary Principle FAQs," at http://www.sehn.org/ppfaqs.html.) The key idea is that scientific certainty should not be a precondition for taking measures to protect human health and the environment.

27 World Bank Project Brief Number GE-PO-79865, "Capacity Building for the Implementation of the Cartagena Protocol," March 27, 2003.

28 *Ibid.*, pp. 22-25.

29 *Ibid.*, p. 17.

30 *Ibid.*, p. 19.

31 *Ibid.*, p. 36.

32 Review by Dr. Klaus Ammann, *ibid.*, pp. 39-40.

33 "Who We Are: History of the CGIAR," at http://www.cgiar.org/who/wwa_history.html.

34 Statement by CGIAR Chairman Ismail Serageldin, at http://www.worldbank.org/html/cgiar/publications/mtm97/isopen.pdf.

35 CGIAR, "Who We Are," *op.cit.*

36 *Ibid.*

37 A simple early experiment with pigment genes in petunias demonstrated, for example, that genetic engineering can result in the silencing of neighboring genes: rather than producing brighter flowers by doubling the amount of pigment gene, researchers found that many of the resulting flowers had no color at all. See Ricarda Steinbrecher, "Ecological Consequences of Genetic Engineering," in Brian Tokar, ed., *Redesigning Life? The Worldwide Challenge to Genetic Engineering*, London: Zed Books, 2001. See also Mae-wan Ho, "Unstable Transgenic Lines Illegal," London: Institute for Science in Society, March 2003, at http://www.i-sis.org.uk/UTLI.php.

38 William Lesser, *The CGIAR at 31: An Independent MetaEvaluation of the Consultative Group on International Agricultural Research: Thematic Working Paper: Review of Biotechnology, Genetic Resource, and Intellectual Property Rights Programs*, Washington, DC: The World Bank, June 30, 2003.

39 *Ibid.* p. 17; also Derek Byerlee and Ken Fischer, *Accessing Modern Science: Policy and Institutional Options for Agricultural Biotechnology in Developing Countries*, World Bank Agricultural Knowledge and Information Systems Thematic Team, November 3, 2000.

40 *Ibid.*, p. 20.

41 *Ibid.*, p. 54.

42 Michael Pollan, "The Great Yellow Hype," *New York Times Magazine*, March 4, 2001. Pollan cites the finding, confirmed by other independent analysts, that a child would have to eat 10 to 15 pounds a day to meet his or her requirement for vitamin A, and would still require a sufficiently functioning fat metabolism to transform beta carotene into vitamin A, a serious obstacle for those experiencing malnutrition. A wide array of cheaper, more sustainable solutions to vitamin A deficiency is readily available. See, for example, Vandana Shiva,

"Genetically Engineered Vitamin A Rice: A Blind Approach to Blindness Prevention," in Brian Tokar, ed., *Redesigning Life? The Worldwide Challenge to Genetic Engineering*, London: Zed Books, 2001; also see Chapter 7 of this book.

43 Lesser, *The CGIAR at 31, op.cit.*, p. 19.

44 *Ibid.*

45 Byerlee and Fischer, *Accessing Modern Science, op.cit.*, p. 23.

46 Lesser, *The CGIAR at 31, op.cit.*, p. 5.

47 Devinder Sharma, "Bill Gates' Rescue Package," ZNet Commentary, November 5, 2003, at http://www.znet.org.

48 Lesser, *The CGIAR at 31, op.cit.*, pp. 51-56.

49 "ISAAA in brief" at http://www.isaaa.org/Flyer.htm.

50 Genetic Resources Action International (GRAIN), "ISAAA in Asia," *op.cit.*, ref. 16.

51 "ISAAA in brief," *op.cit.*

52 "ISAAA in Asia," *op.cit.*

53 *Ibid.*

54 Peter Shinkle, "Monsanto reaps some anger with hard line on reusing seed," *St. Louis Post-Dispatch*, May 12, 2003.

55 "Investing in Destruction — The World Bank and Biodiversity," *op.cit.* On the myths of "carbon offset forestry," see Larry Lohmann, "The Dyson Effect: Carbon 'Offset' Forestry and the Privatization of the Atmosphere," *Corner House Briefing No. 15*, Dorset, UK: The Corner House, July 1999.

56 Conservation Finance Alliance, *Conservation Finance Guide: Introduction*, pp. 4-5, at http://guide.conservationfinance.org.

57 On the problematic nature of debt-for-nature swaps, see Brian Tokar, *Earth for Sale: Reclaiming Ecology in the Age of Corporate Greenwash*, Boston: South End Press, 1997, pp. 165-168.

58 *Conservation Finance Guide: Introduction*, p. 2 On "carbon investments" see note 55 above.

59 Brian Tokar, *Earth for Sale*, esp. — op. cit., Chapter 8; Vandana Shiva, "Biopiracy: The Theft of Knowledge and Resources," in Brian Tokar, ed., *Redesigning Life?*; Bill Weinberg, "The Battle for Montes Azules: Conservation as Counterinsurgency in the Chiapas Rainforest," *Native Americas*, Spring 2003, pp. 40-53.

60 "Universe of BEFs," at http://guide.conservationfinance.org.

CHAPTER 4

1 Mae-Wan Ho, *Genetic Engineering: Dreams or Nightmares?* New Delhi: Research Foundation for Science, Technology and Ecology, 1997, p. 10.

2 Quoted in "Genetic Engineering: A Cautionary Approach," Fairfield, Iowa: Institute of Science, Technology and Public Policy, at http://www.istpp.org/genetic_engineering.html.

3 A 2004 study published in the British journal *Nature* demonstrated that transgenic DNA does indeed survive passage through the stomach and small intestine, and can transfer into the microflora of the small bowel. Trudy Netherwood,

et al., "Assessing the survival of transgenic plant DNA in the human gastrointestinal tract," *Nature,* Vol. 22, February 2004, pp. 204-209.

4 R.A. Ennos, "The influence of agriculture on genetic biodiversity," in *Biodiversity and Conservation in Agriculture,* British Crop Protection Council Symposium Proceedings, Vol. 69, 1997, pp. 15-23.

CHAPTER 5

1 Adam Bradbury, "GM won't cure hunger in Africa," Friends of the Earth: *Real Food News* (on-line digest), January 2002.

2 Rick Weiss, "US on GE-Tainted Food Aid — 'Beggars Can't be Choosers'," *Washington Post,* July 31, 2002.

3 Philip McMichael, *"Global Food Politics,"* In: *Hungry for Profits: The Agribusiness Threat to Farmers, Food and the Environment,* Fred Magdoff, John Bellamy Foster, and Frederick H. Buttel, eds., New York, Monthly Review Press: 2000.

4 Edward Clay and Olav Stokke, *The Changing Role of Food Aid and Finance for Food,* In: *Food Aid and Human Security,* Edward Clay and Olav Stokke Frank, Eds., Cass Publishers: London and Portland, 2000.

5 *Ibid.*

6 Richard Ball and Christopher Johnson, "Political, Economic and Humanitarian Motivations for PL 480 Food Aid: Evidence from Africa," *Economic Development and Cultural Change,* Vol. 44, No. 3, April 1996.

7 Susan George, *How the Other Half Dies: The Real Reasons for World Hunger,* New Jersey: Rowman and Littlefield Publishers, 1977.

8 Evelyne Hong, "Globalization and the Impact on Health: A Third World View," prepared for: The Peoples' Health Assembly, December 4-8, 2000, Savar, Bangladesh.

9 Press release sent by the KMP Filipino Farmers Union, April 14, 2001.

10 Edward Clay and Charlotte Benson, *Acquisition of Commodities in Developing Countries for Food Aid,* Baltimore: Johns Hopkins University Press, 1993.

11 Unattributed sidebar in Vernon W. Ruttan, ed., *Why food aid?,* Baltimore: Johns Hopkins University Press 1993.

12 Fred Pearce, "UN is slipping modified food into aid," *New Scientist,* September 19, 2002.

13 Norfolk Genetic Information Network, "How the US Violates the Food Aid Convention," 2002, at http://ngin.tripod.com/forcefeed.htm.

14 Food First, "Food Aid in the New Millennium: Genetically modified food and foreign assistance," Food First Fact Sheet, 2001 at http://foodfirst.org/pubs/factsheet.2000/biotechfs1.html.

15 "Genetically modified food aid causes outrage," *Afrol News,* June 15, 2002, at http://www.afrol.com/News2002/afr016_foodaid_gmo.htm.

16 Matt Mellen, "Who is getting fed?" *Seedling* (GRAIN publications), April 2003.

17 Naomi Klein, "When Choice Becomes Just a Memory," *The Guardian,* Thursday, June 21, 2001.

18 From USAID's website: http://www.usaid.gov.

19 Greenpeace, *USAID and GM Food Aid,* October, 2002, http://www.greenpeace.

org.uk.

20 *Ibid.*

21 Devlin Kuyek, *ISAAA in Asia: Promoting corporate profits in the name of the poor,* October 2000.

22 Personal communication.

23 Matt Mellen, "Who is getting fed?," *op. cit.*

24 *Ibid.*

25 *Seeds of Doubt: North American Farmers' Experiences of GM Crops,* Bristol, UK: Soil Association, 2002.

26 Michael Passoff, *Special Report: Genetically Engineered Food and Financial Risk,* As You Sow Foundation, San Francisco: May, 2001.

27 GRAIN, "Sprouting Up: Global growth of GM crops: Success or failure?" *Seedling* (GRAIN publications), January 2003.

28 Michael Passoff, *Genetically Engineered Food and Financial Risk, op. cit.*

29 *Ibid.*

30 Food First, "Food Aid in the New Millennium," *op. cit.*

31 Norfolk Genetic Information Network, "Foreign Aid Bill Boosts Biotech Funding," 2000.

32 International Center for Trade and Sustainable Development, "GMOs continue to cause a stir," July 18, 2000, at http://www.ictsd.org.

33 Food First Press Release, "Eastern Europe Opposes US Budget Proposals to Push Agricultural Biotech, July 11, 2000, http://www.foodfirst.org/media/news/2000/7-11eeurope.html.

34 Rory Carroll, "Zambians starve as food aid lies rejected," *The Guardian,* Thursday, October 17, 2002.

35 "Remarks by the President at the BIO 2003 Convention Center and Exhibition," June 23, 2003, from http://www.whitehouse.gov.

36 Adam Bradbury, "GM won't cure hunger in Africa," *op. cit.*

37 *Ibid.*

38 *Ibid.*

39 *Ibid.*

40 C.S. Prakash and Gregory Conko, "Scientists Applaud Limited Activist Support for GM Food Aid," *AgBio World,* September 5, 2002, at http://www.agbioworld.org.

41 Fulvio Grandlin, "Seed Security for Africa's Farmers," *Seedling* (GRAIN publications), October 2002.

42 Shapi Shacinda, "U.S. Needs Good Plan to Give AIDS Funds — Health Chief," Reuters, December 1, 2003.

43 "US cuts off food aid to Sudan over GMOs," Norfolk (UK) Genetic Information Network *GM Watch Daily,* March 15, 2004, at http://www.gmwatch.org.

44 Greenpeace, *USAID and GM Food Aid, op. cit.*

45 Mark Duffield, "The Political Economy of Internal War: Asset Transfer, Complex Emergencies and International Aid," p. 59, In: *War and Hunger: Rethinking International Responses to Complex Emergencies,* Joanna Macrae and Anthony Zwi, with Mark Duffield and Hugo Slim, Eds., Save the Children, UK, London and New Jersey: Zed Books, 1994.

46 *Ibid.*, p.59.

47 *Ibid.*, p. 60.

48 *Ibid.*, p. 59.

49 Action Alert, Research Foundation for Science, Technology and Ecology, June, 2000.

50 C.S. Prakash and Gregory Conko, "Scientists Applaud Limited Activist Support for GM Food Aid," *op. cit.*

51 Elijah Rusike, "Izwi neTarisiro : Zimbabwe's Citizen Jury," *Seedling* (GRAIN publications), October 2003.

52 Quoted in Susan George, *How the Other Half Dies, op. cit.*

53 Adam Bradbury, "GM won't cure hunger in Africa," *op. cit.*

CHAPTER 6

1 David Quist and Ignacio Chapela, "Transgenic DNA introgressed into traditional maize landraces in Oaxaca, Mexico," *Nature*, Vol. 414, pp. 541-543, 2001.

2 "Zambia to export over 50,000 tonnes of surplus maize: official," Agence France Presse, December 13, 2003, archived at http://www.gmwatch.org/archive2. asp?arcid=1887. Nonetheless, hunger was still reported during the Summer of 2003, with growing food insecurity in southern and western provinces of Zambia; several southern districts were declared disaster areas. The increase in domestic corn production in 2003 was attributed to government subsidies for subsistence farmers, improved rainfall, and the efforts by the government and NGOs to repair and create irrigation systems, increase planting area, and provide farmers with high-quality, drought-resistant seeds.

CHAPTER 7

1 Recent statements by supporters of the AgBioWorld Foundation and its founder, Dr. C.S. Prakash, are available at http://www.agbioworld.org.

2 UN Food and Agriculture Organization, "Rome Declaration on World Food Security, and World Food Summit Plan of Action," Rome, 1996.

3 See, for example, papers circulated by the biennial World Agricultural Forum, sponsored by several leading agribusiness, biotechnology and public relations firms, at http://www.worldagforum.org.

4 United Nations Development Program, *Human Development Report 2001: Making Technology Work for Human Development*, New York: Oxford University Press, 2001.

5 *Ibid.*

6 *Ibid.*

7 See, for example, "Distress: A way of life in Kalahandi," *The Hindu*, New Delhi, May 13, 2001, at http://www.hindu.com/thehindu/2001/05/13/stories/13130610.htm.

8 "Transgenics: US Team meets Chief Justice of India," *The Hindu*, New Delhi, January 5, 2001.

9 Presentation by Dr. Adrian C. Dubock of Zeneca Plant Science (now Syngenta) at a conference, "Sustainable Agriculture in the New Millenium: The impact of biotechnology on developing countries," Brussels, May 28-31, 2000.

About the Authors

AZIZ CHOUDRY is an activist, researcher, and writer. He has been an organizer for GATT Watchdog, based in Aotearoa (New Zealand), and active for many years in struggles for indigenous peoples' sovereignty and against capitalist globalization. He is a regular commentator for ZNet (www.zmag.org) and his writing is published widely in Asia, the Pacific, and North America. He is member of the board of convenors of the Asia-Pacific Research Network (www.aprnet.org), the board of the Global Justice Ecology Project, and the advisory board of the Action, Research and Education Network of Aotearoa (ARENA). He is editor of the book, *Effective Strategies in Confronting Transnational Corporations* (Asia-Pacific Research Network, Manila, 2003), and is currently an Associate Fellow of McGill University's Centre for Developing-Area Studies.

S'RA DESANTIS is a member of the Diggers Mirth Collective Farm, raising organic vegetables in Burlington, Vermont, and a co-founder of the Vermont Mobilization for Global Justice. She is the author of *Freedoms That Are Abolished: FTAA Green Paper* (for ACERCA: Action for Community and Ecology in the Regions of Central America, 2001) and *Control Through Contamination: US Forcing GMO Corn and Free Trade on Mexico & Central America* (ISE Biotechnology Project, 2003). DeSantis has published articles in *Z Magazine* and *Synthesis/Regeneration* on the impacts of GE corn in Mexico, and spent two months in Nicaragua and El Salvador in 2003 presenting workshops on the risks of genetic engineering to farmers and academics in the region.

MWANANYANDA MBIKUSITA LEWANIKA is trained as a biochemist and is currently executive director of the National Institute for Scientific and Industrial Research in Lusaka, Zambia. He was a key figure in the Zambian government's effort to investigate the implications of GE food aid, serving on the scientific mission that traveled to the US in 2003, and ultimately urged their government to continue refusing GE corn. Lewanika was Zambia's chief negotiator for the Cartagena Protocol on Biosafety, and is also active in the preservation of traditional knowledge, representing the Southern

African Traditional Leaders' Council for the Management of Natural Resources at various international meetings. His account of the GE food aid debate in Zambia was prepared for the *Biodevastation 7* conference in St. Louis, May 16, 2003.

SHIRI PASTERNAK is a member of the Food Not Lawns collective, a network of researchers, farmers, artists, and community organizers originally based in Montreal. She is a research associate at the Centre for Studies on Religion and Society at the University of Victoria, British Columbia, and a graduate student in the interdisciplinary Cultural, Social and Political Thought program. Pasternak is currently preparing a report for Genome BC on the ethical standing of plants, animals, and the environment. She has given numerous workshops linking the corporate globalization of agriculture to peasant and farmer struggles at home and abroad, and her most recent article, "War By Other Means: Pop-Tarts and Geopolitics in the New Agrarian Order," can be found at www.foodnotlawns.org.

DEVINDER SHARMA is a respected analyst and critic of international food and trade policies. Trained as an agricultural scientist, he currently chairs the New Delhi-based Forum for Biotechnology and Food Security. He is the author of *GATT and India: The Politics of Agriculture*; *GATT to WTO: Seeds of Despair*; and *In the Famine Trap*, and served as the development editor of the *Indian Express*, the largest selling English language daily in India. Sharma has served as a visiting fellow at the International Rice Research Institute, the School of Development Studies (University of East Anglia, Norwich, UK), and Cambridge University, and his weekly column appears in newspapers across India and south Asia.

LAWRENCE TSIMESE has served as an information officer and projects director at the Ghana Organic Agriculture Network, based in Kumasi, Ghana. He is trained in agriculture, agro-forestry, organic certification and inspection, and communication, is a Christian minister and an experienced video producer. Tsimese is a prolific writer on sustainable agriculture and pesticide issues, and has presented his work at several international conferences. He recently completed a year-long internship studying ecological land use at the Institute for Social Ecology in Vermont.

BRIAN TOKAR has been an has been an activist, author, and a leading critical voice for ecological activism since the 1970s, and is a faculty member and Biotechnology project director at the Institute for Social Ecology in Plainfield, Vermont. He is the author of *The Green Alternative* (1987, revised 1992) and *Earth for Sale* (1997), and edited *Redesigning Life?*, an international collection on the politics and implications of biotechnology (Zed Books, 2001). Tokar received a 1999 Project Censored award for his investigative history of the Monsanto company (*The Ecologist*, September/October 1998). He has lectured throughout the US, as well as internationally, and his articles on environmental issues, emerging ecological movements, and resistance to genetic engineering have appeared in Z *Magazine*, *The Ecologist*, *Earth Island Journal*, *Wild Matters* (formerly *Food & Water Journal*), *GeneWatch*, *Toward Freedom*, and many other publications. He holds concurrent degrees from MIT in biology and physics, and a masters degree in biophysics from Harvard University.

Index

AATF (African Agricultural Technology Foundation)

ADM (Archer Daniels Midland Corporation) Kraft: 48.

Africa: 3, 5, 12, 14, 41-42, 44, 45, 47-48, 51, 57, 63-64, 67-72, 78-83, 86, 87, 88-89, 91, 94, 96. Burundi: 80. Egypt: 48, 78. Ethiopia: 54, 69, 70, 73, 81-82. Ghana: 14, 67, 69-70. Kenya: 54-55, 60, 64, 68, 69, 78, 82. Malawi: 69, 80, 82. Mozambique: 80, 82. Somalia: 75-76. South Africa: 45, 64, 82, 88, 96. Sudan: 83. Tanzania: 64, 69. Uganda: 64, 69, 82. Zambia: 5, 7, 14, 67, 80-81, 82, 84, 86-90, 113 n.6:2. Zimbabwe: 68, 78, 80, 81, 84.

Africa Model Law on Biosafety: 71.

African Agricultural Technology Foundation (AATF, Kenya): 78-79.

African Diversity Network: 82.

AgBioWorld Foundation: 91.

Agreement on Agriculture (AOA; WTO agreement): 39, 46-47. (See also agriculture)

agribusiness: 4-9, 11-14, 17, 27, 29, 33, 39, 40, 47-48, 57, 66, 71, 75, 77, 85, 91-93. (See also Aventis, Cargill)

Agricultural Genetic Engineering Research Institute (Egypt): 78.

Agricultural Research Institute (Kenya): 78.

allergies: 8, 22-23, 88.

animals: 10, 18, 19, 22, 23, 40-42, 44, 49, 57, 67, 87. amphibians: 21. cattle: 26, 32, 48, 65. chickens: 4. fish: 9, 21, 60. livestock: 13, 29, 53, 60, 61, 76, 95, 98. pigs: 19, 22. rats: 9, 23-24. turkeys: 4.

AOA (Agreement on Agriculture)

apomixis technology (inducing seed formation without pollination): 61.

Asia: 2, 3, 12, 36, 40, 42, 63, 64, 78, 80, 99. (See also India) Bangladesh: 99. China: 79. Indonesia: 54, 64, 78, 99, 108 n.17. Japan: 20, 23, 28, 36, 37, 40, 52, 75-76. Malaysia: 64, 78. Pakistan: 74, 99. Philippines: 60, 64, 76, 78, 80, 96. Singapore: 43. South Korea: 10, 23, 36. Sri Lanka: 47, 80, 108 n.17. Taiwan:

36. Thailand: 64, 78. Uzbekistan: 74. Vietnam: 37, 52, 64, 78, 99.

AusBiotech Alliance (Australia): 55.

Australia: 48, 55.

Aventis CropScience (now Bayer CropScience): 23, 56-57, 91. (See also StarLink)

Avoka, Cletus (Ghana's Minister for Environent, Science and Technology): 70.

bacteria: 18, 20, 22, 69. disease: 9, 22. insecticidal: 18-19, 23. (See also Bt-varieties)

Bayer: 18, 57, 63, 91. (See also Aventis CropScience, Rhone Poulenc Agro)

beef: 4, 48.

Berne Declaration: 46.

beta carotene (vitamin A precursor): 54, 109 n.42.

Biodevastation events: 3, 6.

biodiversity: 13, 25, 27, 33, 40, 42-43, 52, 60, 62, 64-65, 67, 70-71, 82, 88. (See also African Diversity Network, Biodiversity Convention, Biodiversity Enterprise Fund, CONABIO)

Biodiversity Convention: (See Convention on Biological Diversity, UN)

Biodiversity Enterprise Fund (IFC project): 65.

bioprospecting: 65.

biosafety (cf food safety under food production): 56, 57, 67-68, 70-71, 77, 78. (See also Cartagena Protocol)

Biotechnology Industry Organization: 10, 55.

Bové, José (Confédération Paysanne spokesperson): 4, 5.

boycotts: 67.

Brazil: 23.

Bristol-Myers: 39.

Bt (Bacillus thuringiensis, an insecticidal soil bacterium): 18-19, 21, 23, 26, 54, 64, 78, 81. (See also bacteria; corn; crops: cotton, potatoes; Cry9C; Terminator; toxins)

Bush, George W. (US President) and administration: 6-7, 29-30, 48, 50, 81.

CABIO (Collaborative Agriculture Biotechnology Initiative)

CAFTA (Central American Free Trade Agreement)

Calgene: 48.

campesinos (peasant farmers): (See farmers)